UNSCREWED

Salvage and Reuse Motors, Gears, Switches,
and More from Your Old Electronics

Ed Sobey

CHICAGO
REVIEW
PRESS

Library of Congress Cataloging-in-Publication Data
Sobey, Edwin J. C., 1948–
  Unscrewed : salvage and reuse motors, gears, switches, and more from your old electronics / Ed Sobey.
    p. cm.
  Includes index.
  ISBN 978-1-56976-604-0 (pbk.)
  1. Electronic apparatus and appliances—Design and construction—Amateurs' manuals. 2. Electronic apparatus and appliances—Recycling—Amateurs' manuals. 3. Salvage (Waste, etc.)—Amateurs' manuals. I. Title.
  TK9965.S66 2011
  621.3815—dc22

                            2011004953

Cover and interior design: Sarah Olson
Cover image: D-BASE/Stone/Getty Images

Published by Chicago Review Press, Incorporated
814 North Franklin Street
Chicago, Illinois 60610
ISBN 978-1-56976-604-0
Printed in the United States of America

To the City of Akron—thank you for launching the National Invention Center and inviting me to be a part of it.

To invent, you need a good imagination and a pile of junk.

—Thomas Alva Edison

This book will help you with the second part.

CONTENTS

ACKNOWLEDGMENTS

Collecting computers and peripherals for dissection is a challenging task in itself. Thanks go to my network of friends who are computer geeks and techno-users. Michael Meyers of Eastside Computer provided me with a treasure trove of components and some good thoughts on how stuff works. Michael also keeps my computers operating so I can write books. Thank you.

Running friend Carl Kadie, a researcher for Microsoft, gave me some unique treasures, too. I think he also regifted me the computer yoke that I had given him the year before. Anyway, Nancy, his wife, is delighted to be rid of some of the stuff crowding their closets. Norbert Geer contributed a monitor for dissection, and Jerry Gardner brought it to our next run. Thanks, guys.

Richard Rundle provided an LCD monitor, and John Dickson gave me his old TiVo. Mark King not only gave me his scanner/fax/printer, he also brought it to me. Seth Leary loaned me one of his metal detectors, but I haven't returned it.

John Weigant contributed several items he rescued from the Portland Goodwill Outlet, where he often shops for yesterday's treasures. He also built the glove/boot/sock dryer pictured in the fan project. Dave Foley with Oregon Community Communications in Roseburg, Oregon, provided technical assistance and component identification. He also answered many of my questions about circuitry and components. Thanks, Dave.

George Gerpheide was generous with his time in recounting to me the history of his invention, the touchpad.

INTRODUCTION

Let's go on a treasure hunt—let's plunder old gizmos and gadgets for the good parts inside! The hunt will take us deep into the dark recesses where only engineers usually venture as we search for useable components, understanding, and techno-entertainment. If you are curious, if you want to know what's on the inside of that plastic or metal case, if you like to know how things work, this hunt is for you.

What might you find when you open up some piece of inoperative technology? You might discover component parts that you can use: motors, switches, magnets, gears, shafts, pulleys, belts, lenses, and lots of screws. Some of the parts will have no immediate utility but might inspire creative thinking on your part. Your curiosity and creativity, empowered by a pile of parts, might equal something new and cool. Totally Edison!

Operating on a broken device may lead you to find out why the gizmo isn't working, and then maybe you will be able to return it to serviceable duty. For sure, you will find things you didn't expect to find, and you'll uncover things that you can't identify. Like any good treasure hunt, you can't predict exactly what you'll find or exactly how you'll get there.

But better than an X on a map or MapQuest directions, this book will guide you to the treasures. It will steer you away from danger and help you overcome the obstacles of the engineered world.

Rules of the Hunt

In over 30 years of teaching reverse engineering, I have come up with important guidelines to ensure a safe process. All of the rules have stories behind them that I won't tell. Trust me—these rules are good ones to follow.

- **Make sure the device owner agrees to your treasure hunt.** Make no promises to fix the device or even return it in as good a condition as it was delivered to you. If the owner doesn't agree, don't accept the device.

- **Cut off any power cords.** Mistakes with power cords can be serious. Remove cords and bend the prongs outward so that the plugs cannot be inserted into outlets. Toss the cords into the trash where kids won't find them.

- **Protect your eyes.** Your eyes are your most vulnerable components. Springs jump, bits of metal and plastic fly, and tools slip. Wear glasses or, better yet, safety goggles.

- **Pry away from you.** Most take-apart accidents occur when someone is pushing very hard on a screwdriver and it slips. It's okay if it slips—as long as it slips away from you and your buddy.

- **Don't torture that VCR!** No hammering. No retribution by sawing. You can almost always get the components out without beating them up. It's a puzzle. If you're stuck, try a different approach. The person who put the device together didn't whack that component into place, so you shouldn't need to whack it out.

- **Watch out for capacitors.** Most are innocuous, but some pack a serious, toss-you-on-the-floor-and-make-you-scream wallop. Cameras, even disposable ones that have strobes, have high-voltage capacitors. Microwave ovens, refrigerators, and blenders also have capacitors to watch out for. Those in televisions and computer monitors (CRTs) can send you to the Big Take-Apart Lab in the Sky—which is why I don't include them here. This book will alert you to potentially dangerous capacitors in the featured devices and will tell you how to render them harmless.

- **Lefty loosens.** Not in every case, but almost always, turning counterclockwise will get the screw loose.

- **Wash your hands.** Mom was right. After messing around with the exotic materials that make up gizmos and gadgets, wash your hands. Whatever is in them doesn't belong in you.

- **If you don't know what it is, don't cut it.** I sometimes do recommend using a rotary cutting tool—but not to explore blindly by cutting stuff up. Nastiness resides inside certain components—such as circuit boards—and you don't want to release nastiness into the air by removing the protective covering.

Tools You'll Want

A few screwdrivers and a pair of needle-nose pliers with wire cutters will handle most of the projects. Phillips screwdrivers are the workhorses of this treasure hunt. One isn't enough. You will want a variety of sizes from the very small to the medium-large.

Here are a few other tools that can help.

- **Magnifying eyeglasses.** Get a pair of 2x magnifying eyeglasses from the drugstore. It's a lot easier to work with small parts when you can see them.

- **A multimeter.** Electronic multimeters are inexpensive, less than $10. They let you test circuits and switches, measure resistance, and measure voltage. Some measure current and test diodes and transistors. Even a cheap meter will be helpful.

- **A Swiss Army knife.** A Swiss Army knife's Phillips screwdriver fits more screw heads than any other driver, and its knife blades are always useful, as are the scissors. If a Swiss Army knife is always in your pocket, you're always ready to take something apart.

- **Two 9-volt batteries and some alligator clip leads.** With these you can quickly test the motors you extract and the LCD screens you find, and you can do some circuit bending on musical instruments.

- **Additional screwdrivers, pliers, and wrenches.** Jeweler's screwdrivers, flat-nose or slip-joint pliers, Vise-Grip pliers, and Allen wrenches help too. Some devices, such as hard drives, have Torx screws. You can get them open without a Torx driver, but it's much more difficult.

- **A rotary cutting tool.** For getting into a molded plastic case, a rotary cutting tool—for example, a Dremel—is a must. It can also cut through that recalcitrant screw that just won't come out. If a screw head is stripped, you can cut a new notch and then remove the screw with a flathead screwdriver.

⊛ **A digital camera and notepad.** A camera allows you to "remember" the original order of parts, before they got scrambled. A notepad and pen records not only parts numbers to look up on the Internet but also those great ideas you will generate.

The Invention of the Phillips Screw

Phillips screws are everywhere. Who invented them? Henry Phillips of Portland, Oregon, for whom the screws are named, didn't originate or produce them himself, but he took the original design by J. P. Thompson, who held the first patent, and improved upon it. Mr. Phillips's revised design was phenomenally successful, and in six years Phillips screws were made by most American screw manufacturers. The beauty of the screws is that they center and hold the driver in place, not letting it slip, as slotted screws do.

Trash and Treasure

In this modern age, the rate of product turnover is astounding. Thirty years ago your telephone was expected to last a lifetime. Now you replace it every two years when you get a new contract. Your computer is outdated months after you've purchased it. The latest newfangled gadget quickly becomes a historical artifact.

The good side of all this is that there is a lot of stuff being tossed away—and a lot of it is interesting to peer into. So where can you find ammunition for your take-apart hobby? Friends and family probably have old appliances waiting for disposal or thrift stores. Goodwill Outlet stores sell nonworking devices for as little as 99 cents each. Other thrift stores may be willing to hand over the stuff they receive but can't sell. Of course, garage sales and thrift stores will also sell you (hopefully) working appliances, but these are more expensive and potentially useful for their original purpose. Free or nearly free is best.

Some of the components you remove need special handling. Please dispose of hazardous materials with the environment and local laws in mind. Yes, you can probably sneak them into your garbage can, but when you do, you degrade your own environment and waste useful materials. Batteries, transformers, cathode ray tubes, and computers should be recycled and not put into the waste stream. Check online for the nearest recycling and disposal center that handles each component. For batteries, check out www .ehso.com/ehshome/batteries.php#Summary to guide you in disposal.

Your Guide to the Unscrewed World

Enough chatter. This book is geared to help you do cool stuff. Each item in the book is one I recommend you try to disassemble. Some are more rewarding than others, however, which I indicate with the Unscrewed Value Index, or UVI. The UVI has three elements: value of the parts you can recover, fun/ discovery value of taking the device apart, and the negative cost of disposal. If an item requires special handling in disposal, that reduces its UVI.

And one final admonishment: Every part you extract was designed by an engineer and included for a specific purpose. The accountants at the manufacturing facility were probably yelling, "Do we *really* need that part?" But there is a reason it's there. Can you figure it out? As you dig your way through that VCR, appreciate the aesthetics of design, the economy of arrangement, and the manufacturing . . . and be inspired by them!

AUDIOCASSETTE PLAYER

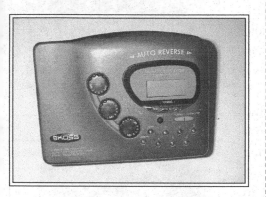

Treasure Cache
DC motor
Drive belts and pulleys
Gears
LCD
Magnetic tape head
Piezo speaker
Rollers/wheels
Springs

Tools Required
Flathead screwdriver
Phillips screwdriver
Scissors

This portable entertainment system was amazing in its day. Sony launched a revolution in 1979 when it introduced the Walkman. The model taken apart here is a later model from a rival company. Today, most cell phones have more audio features than a Walkman, and can play more songs. But one of the wonderful things about this older technology is that it houses so many useable mechanical parts.

Lefty Loosens

Four Phillips screws hold the back
panel to the rest of the player.
Underneath the panel is the back
of a circuit board; a few screws hold
this board to the front frame. On
the right are a motor, two pulleys,
and a belt.

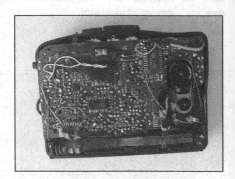

I lifted the circuit board off and
cut the several wires connecting it
to the rest of the player. In cutting
wires it's always good to leave as
much wire as possible attached to
any components you might want to
use later.

The pulley on the motor shaft
belt drives a double pulley above it.
This double pulley drives a second
belt that powers the two white pulleys. Pressing either the "Fast Forward" or "Reverse" buttons on the
top of the player moves one of these
white pulleys and its associated
white gears. When one of these pulleys moves, it takes up the slack in
the large belt, which is how it spins
without turning the other white
pulley.

Between the two large white pulleys is the switch for the "Reverse"
button. Pressing this button pushes
together the contacts to power the
circuit as well as move the pulley
into place.

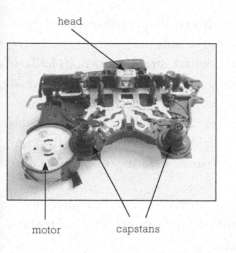

head

motor   capstans

A few screws hold plastic retaining frames over the pulleys and the motor. These come out, freeing the belts, motor, pulleys, and gears. I used a flathead screwdriver to pry them out.

The other side of this assembly is where the tape moves. The magnetic head is here. Depressing the "Play" button moves the head onto the tape so it can read the magnetic fields of the recordings. Guide rollers guide the magnetic tape, and springs apply tension to the tape. The capstans hold the cassette in place and allow the tape to wind and unwind.

The top cover pivots open on plastic hinges so you can drop in a cassette. The plastic housing for the cassette is held to the top cover by a few screws. There is a circuit board in the top that holds the control switches and an LCD screen. Three of the switches, on the left side, are variable resistors. The others are rubber dome contact switches. Wires running from the board go to a piezo speaker that is lightly glued to the inside of the top cover. This can't produce high-quality sounds, so this player is designed to be used with earphones.

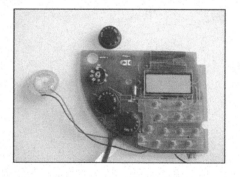

What Now?

The plastic case itself could be an interesting container for small parts. You would want to cut out much of the plastic inside to leave as large a space as possible and reassemble the body. You could access the parts stored inside through the hinged top where the cassette would go.

The rollers could become wheels for a tiny vehicle. The motor, which runs on 2 volts, could drive such a vehicle with the belts.

Try messing around with the LCD. Connect various pairs of its leads to a 9-volt battery to see if you can get it to make numbers or letters. It would be interesting to test the heads to see if they put out enough voltage to generate sounds. You would have to pass the signal through an amplifier (Radio Shack sells tiny amplifier/speakers that might work nicely for this). Then run the magnetic strip on your credit card past the head so you can hear your account number.

The piezo speaker could become a pick-up head for a musical instrument, like a homemade electric guitar. Run the leads into an amplifier and try taping the speaker against the bridge of a guitar.

BAR CODE SCANNER

Treasure Cache

Charge-coupled device (CCD)

Magnetic transducer

Metal weight

Tiny lens and mirror

Tools Required

Phillips screwdriver

Rotary cutting tool

Y ou've used these devices hundreds of times at the checkout stand at the grocery or to identify you as a proper card-holding patron. Now's your chance to open one up and, in the process of scientific discovery, render it impotent.

Lefty Loosens

The scanner's plastic bottom comes off with the removal of four screws. This reveals a large metal weight on one side that helps hold the scanner in place on a countertop.

All the action is on the other side of the scan slot. A narrow window in the slot wall allows the light reflected off the bar code to be read inside the case, while keeping ambient light out. Two circuit boards hold the electronics. A circular array of LEDs illuminates the card: they shine through the window and reflect off the white surfaces in the code. The reflected light comes back through the window and through the center of the circular array. It reflects off an angled mirror, through a lens, and onto a detector encased in a brass housing at the far end. The brass housing has a narrow slot to admit only the light from one white bar at a time.

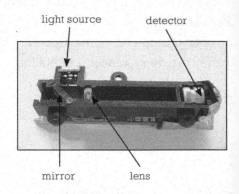

light source   detector

mirror   lens

To get to the sensor inside the brass housing, I used a Dremel rotary cutting tool. Inside is a tiny sensor encased in a plastic envelope. The sensor appears to be a charge-coupled device, the same technology used in digital cameras. Other bar code scanners use photo-diodes to read the bars.

This scanner has a small magnetic transducer to signal the user with a sound that a card has been read or not. Inside the transducer is a coil of fine copper wire wrapped around a metallic core. This sits inside a circular or toroid magnet. A metal disk is held in place by the magnet. As electronic pulses pass through the windings, they create a magnetic field that flexes the sound-making metal disk up and down. This movement generates sound waves that tell you, for instance, "Oops, your card didn't get read."

magnet

What Do Those Bars Mean?

The bar codes on most items you buy use a numbering system called UPC, for Universal Product Code. Each UPC bar code has 12 digits plus start, end, and middle bit patterns. The first six digits identify the manufacturer, and the second six identify the product. The manufacturer assigns a number for each product, each type of package, and each size using five of the six digits. The last digit is a checksum digit. The computer running the bar code scanner adds up combinations of the code numbers and performs mathematical calculations with them, finally comparing the result with the sixth digit. If the computer doesn't come up with the same digit, it alerts the operator that the code was incorrectly read.

What Now?

If you haven't destroyed it, the magnetic transducer might be fun to experiment with. What sounds can you generate with it? If you wire it into a working electronic keyboard or guitar, does it make sounds like the dynamic speaker?

BUBBLE GUN

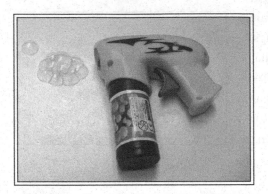

Treasure Cache
DC motor
Gears
Pump
Spring
Switch
Valve/steel ball

Tools Required
Phillips screwdriver
Towel

When you pull the trigger on a bubble gun, it lifts the applicator to cover the end of the nozzle with bubble juice, and when the fan starts, bubbles are blown out. Pulling the trigger also energizes the electric motor that both blows air through the nozzle and pumps bubble juice from the reservoir. It's a good toy, but a better take-apart. The engineering inside is inspiring. One motor performs two functions that require different rotational speeds, so gears are employed. Before you take the gun apart, insert a set of new batteries and a fresh bottle of bubble solution to give it a try; then ask yourself how you would make a device that could blow bubbles.

Lefty Loosens

Four visible screws hold the two halves of the bubble gun together. Two more lurk beneath the two AA batteries inside the battery case. (After removing any screws inside the battery case, you should reinsert the batteries so you can activate the gun while it's disassembled.)

Lay the bubble gun on a durable surface with the battery-case side down. (Bubble juice is bound to leak out, so you'll want to grab a towel too.) Lift off the top half. You can see that the motor drives both a centrifugal fan above and the gears below. Exhaust from the fan is forced down the plastic ducting toward the nozzle.

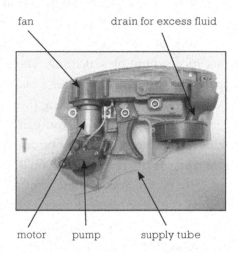

fan   drain for excess fluid

motor   pump   supply tube

At the end of the nozzle is an applicator that moves up and down to spread bubble juice in front of the stream of air. This mimics how you would blow bubbles with your mouth. You dip the wand in solution and then blow air through the film of solution stretched across the circular opening. The applicator draws a sheet of solution in front of the nozzle, and then the fan pumps air out.

Excess juice drains down a drip rod into a funnel, past a check valve (you can see the shiny metal ball in the tube), back into the bottle of bubble solution.

Pushing on the trigger compresses a return spring and closes a simple switch. The switch is a piece of metal (a conductor) that the trigger forces onto the metal case of the motor, thus completing the circuit and energizing the motor. The

motor turns the fan blades at high speed, spinning air in a circle and forcing it outward into the duct inside the barrel. (This is a *centrifugal pump* since it forces air outward.) The trigger also lifts the applicator at the end of the nozzle.

The shaft coming out of the bottom of the motor has a worm gear. This allows the motion to change direction 90 degrees and engage a set of gears that slows the motion and increases its torque. The end of this gearing is a peristaltic pump—it works like your throat swallowing a bite of Cheerios—that moves fluid by squeezing. The pump, operating at a greatly reduced speed due to gearing, squeezes a plastic tube that draws bubble solution.

Unscrew two screws holding the gears to the gun body so you can remove the gears. Lift the bottom gear and on the reverse side you'll see two plastic nubs. These press against the clear plastic tubing. As the nubs turn, they pump bubble juice along the tube from the reservoir to the barrel.

The Invention of the Bubble Gun

Robert DeMars invented the electric bubble gun. Not content with this major impact on American society, he also invented the clip-on handle for a beer can, the resealable flip-top for aluminum cans, and a sleeping bag that's shaped like a giant teddy bear.

What Now?

Reassembling the pump would allow you to make a tiny fountain or water-fall, though it would be more of a watertrickle than a waterfall. Or it could refill your beverage glass with the touch of a button—and you'd be the envy of all your friends.

The pump could also irrigate your indoor plants. Drive it with a micro-processor like a Stamp chip that has a soil moisture (conductivity) probe. While you're away on a trip, your pump could be activated when the soil becomes dry or when you signal it via the Internet.

CAMERA (DIGITAL)

Treasure Cache

Buttons

Charge-coupled device (CCD)

DC motors

LEDs

Lenses

Metal shafts

Polarizing filters

Springs

Strobe light

Tools Required

Jeweler's screwdrivers

Scissors

Talk about a revolution in technology—from millions and millions of rolls of film used each year to nearly none.

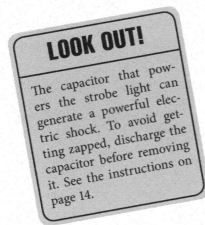

LOOK OUT!

The capacitor that powers the strobe light can generate a powerful electric shock. To avoid getting zapped, discharge the capacitor before removing it. See the instructions on page 14.

Lefty Loosens

Cameras can be difficult to open, but this one required only a small Phillips screwdriver to separate the front and back halves. I cut the conductors on plastic strips running between the two halves.

The back half houses several control buttons, each held in place by small screws. The four-way mode switch allows users to pick four different menus by pushing the round switch cover in any of four directions. Beneath the cover there are three layers of plastic material. One holds four conductive tabs that, when pushed, connect two sides of the circuit below. A panel with four buttons from the lower left corner of the back half uses the same approach for entering commands. The "W" (for wide) and "T" (for telephoto) buttons, along with their electrical connections, pull out. The scene selector knob on the top of the camera pops out. Inside is a pair of metal contacts that you rotate to make connection between pairs of contacts below.

Looking at the back of the front half of the camera, the display window is the dominant component. This lifts out and can be taken apart. Inside are all the components you find in a computer LCD screen, but in much smaller sizes. Two tiny LEDs provide the light source, and several layers of plastic material provide the polar-
ization required. If you're interested in seeing how an LCD works, tak-
ing apart a laptop screen (see page 115) would be better, as the larger size makes it easier to see things.

Beneath the display window are several interesting components. The large battery lookalike cylinder is a capacitor. The capacitor is there to power the strobe light that sits above it. If this capacitor is charged, it packs a wallop—one that you don't want zapping your fingers. To make sure it isn't charged, *while holding the wooden, plastic, or rubber handle of the screwdriver (not the metal!)*, put the blade of a screwdriver on both contact legs of the capacitor at the same time. If it's charged, you'll see a spark. Touch it several times to make sure the charge is dissipated. Disposable cameras have similar capacitors, and even these pack a punch. The strobe requires a sudden burst of electric power at a high voltage, and the capacitors store that power and provide it quickly to the strobe.

connecting pins

Off to the side is the slot that the memory card slides into. This camera didn't have a memory card in it, but the connecting pins are visible.

The chip that transforms light into electric charges is now visible in the center of the camera, directly behind the lens. Like most digital cameras, this one uses a CCD, or charge-coupled device. This is an array of tiny diodes that capture light and capacitors that accumulate charge according to how much light hits their location. The charge is measured and stored as a digital value that can be converted into an image. Each diode in the array is about 3 to 8 microns in size, or about the size of a red blood cell.

The lenses offer an entirely new opportunity for components. Unlike the lenses of film cameras, these have motors in them. One motor opens the diaphragm in front of the lens to allow you to take pictures, another focuses

the lens, and a third provides wide angle and telephoto zoom capabilities. The lens moves in or out to change its focal length.

Inexpensive cameras have fixed focal length lenses: you point and shoot. Some cameras offer "digital zoom": rather than moving the lens out to get images farther away, the camera uses only the image in the very center of the CCD. It blows this image up to full size—a digital sleight of hand that often disappoints.

The eyepiece has a tiny motor as well. The motor turns a shaft that rotates a plastic cam. It moves both lenses, pushing them together or apart. This action is opposed by tiny springs. It's really cool, and you have to wonder how much design thinking went into this one tiny piece. The autofocus is probably provided by an infrared LED mounted on the front of the camera.

Be sure to dispose of the battery properly.

What Now?

The three tiny motors could be powered by solar cells. They could drive very small, solar-powered kinetic sculptures. The autofocus motor and gearing are intriguing. To be useful, this component might have to stay in the camera body, but the wires could come out and be connected to a switch and power.

The Invention of the Digital Camera

Working for Kodak in 1974, Steven Sasson was given a CCD (charge-coupled device) and told to experiment with it. With no real objective in mind, he thought that he would like to see if he could use it to take photographs. It took a year to build a toaster-sized camera that weighed over eight pounds and took fuzzy photos. He quickly corrected the fuzziness, yet ran into the problem of how to display the images.

Computers weren't common in the mid-1970s, so he displayed them on a television. No one could imagine that people would want to take photos to show on their TVs, so not much happened until personal computers came on the market. The first Kodak digital camera was released in conjunction with Apple Computer in 1994; however, others had beaten Kodak to the punch: the first digital camera for consumers had been introduced in 1990. Ten years later, the rush to digital cameras was in full stride.

CAMERA (FILM)

Treasure Cache

   Capacitor

   DC motor

   Gears

   Lens

   Strobe light

Tools Required

   Phillips screwdriver
      (very small) or jeweler's
      screwdriver

   Scissors

Now that they've been replaced by digital cameras, film cameras are available for pennies at thrift stores and garage sales. With a motor to move film from one side of the camera to the other, these prove even more interesting to take apart than digital cameras.

LOOK OUT!

The capacitor that powers the strobe can carry a powerful charge. Be sure to discharge the capacitor before removing it. See details on page 18.

Lefty Loosens

I took apart a Kodak film camera that had a plastic body composed of two parts, front and back. First I opened the battery compartment and removed two batteries. Two tiny Phillips screws on the top were the most visible ones.

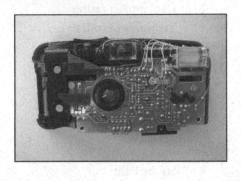

Three other tiny screws became visible once the film door in the back was opened. After I removed those screws, the two halves came apart.

Removing the front half of the camera body revealed the viewfinding lens, the strobe (for flash photography), the camera lens, and a circuit board held in place by a few more tiny screws.

From the top, the film exposure counter and an LED (facing forward) are visible. From the bottom, the very large (330-volt) capacitor is visible. This is the only hazard. If the capacitor is charged and you simultaneously touch both terminals, you will get a nasty wake-up shock. To be sure you don't, *hold a screwdriver by the wooden, plastic, or rubber handle* and touch the screwdriver's metal shaft to both contacts (or legs) of the capacitor at the same time. Having given you this warning, I will admit that it is unlikely that the capacitor you encounter in an old camera will be charged, but I still always take the time to make sure by touching the legs with a screwdriver.

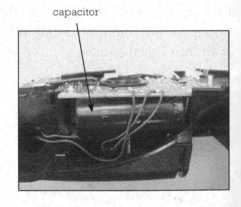

capacitor

The capacitor is mounted on a circuit board. Also located on the board are three diodes, three transistors, three smaller capacitors (no need to worry about discharging them), and a transformer to boost the voltage high enough to charge the big capacitor. This circuit board converts 3.0 volts, from the two batteries, to 330 volts, to charge the capacitor. To do that, it first has to convert the direct current into alternating current for the transformer to work.

Disposable film cameras also have strobes, capacitors, and batteries. Their capacitors are equally powerful, so make sure they are discharged. In the ones I have taken apart, the capacitor is located near the strobe.

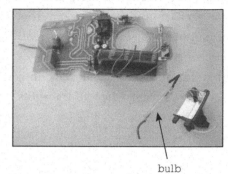

bulb

The strobe assembly lifts off the top of the camera. To free it from its circuit board, I cut the wires. The bulb is held in the strobe assembly by a tiny rubber band. I cut that and the bulb fell out.

Above the viewfinder is a plastic assembly held in place by more tiny screws. Taking them out frees the assembly. The viewfinder lens pulls out of its slots in the camera body.

A few other screws hold a cover in place over the right side of the camera. Under the cover are a series of colorful gears. At the far right side of the gear train is a metal gear. This is mounted on the shaft of a small electric motor. The motor advances the film and rewinds it back into the film canister when all the shots have been taken.

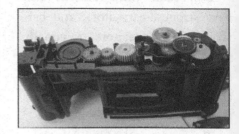

Two screws on either side of the metal gear hold the motor in place. With the screws removed, the take-up roller, with the motor inside, can come out. The motor in this camera still works.

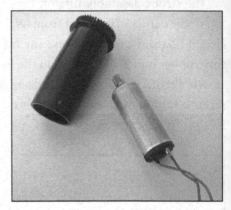

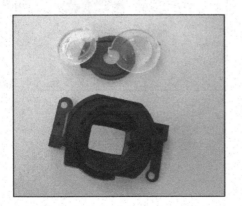

From the back of the camera, the lens and aperture, which sit inside a plastic assembly held in place by two screws, can be removed.

What Now?

The small motor won't deliver a lot of power but would be great for moving light loads. It could be useful in solar-powered applications, as it requires low voltage. Or it would make a good vibrating motor, like the ones in cell phones, if you just add an off-center weight to the motor shaft. It could also be a tiny wind generator to recharge AA batteries: add a propeller to the motor shaft and connect the motor terminals to a battery. Be sure to connect them so the positive current from the motor goes to the positive (+) side of the battery.

The tiny lenses could be great viewing ports for miniature dioramas or models. The gears could be put to use in small automata—motor-powered artistic gizmos.

CD-ROM

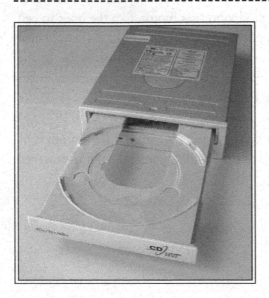

Treasure Cache
   DC motors
   Drive belt and pulley
   Gears
   Laser and assembly
   Lens and mirrors
   Metal cover
   Metal rods/shafts

Tools Required
   Flathead screwdriver
   Phillips screwdriver
   Scissors

Making the switch from recording information magnetically to recording it optically revolutionized data storage, computing, and the music industry. A CD can hold about 480 times as much data as a 3.5-inch floppy disk (700 megabytes of data, compared to 1.44 megabytes). The device that makes this miracle of miniaturization possible is the laser. Only a few decades ago pundits were calling the laser an invention without a use, and now lasers are in homes, stores, and businesses everywhere.

Lefty Loosens

Taking out a few Phillips screws and prying the edges with a flathead screwdriver were enough to get the metal cover off. The underside is a forest of chips and other components with large ribbon connectors stretching like superhighway overpasses.

On the front side, with the faceplate removed, the emergency eject system is revealed. Maybe you didn't know this existed, but if the CD is stuck inside and won't eject, you can poke a straightened paper clip into the tiny hole in the faceplate. The paper clip pushes a plastic piece that rotates to release the CD. Now you can see this mechanism.

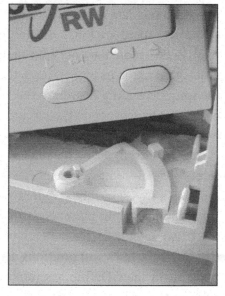

With the cover removed we can see the black circular drive spindle. The brushless drive motor sits beneath this. Not having brushes allows it to operate for long periods of time and to operate without generating the electromagnetic noise that a DC motor makes. Brushless motors also are more efficient and generate less waste heat.

motor with worm gear to open the tray
motor to position the laser assembly
laser assembly

spindle motor

The spindle motor has to spin at different speeds, from about 200 to 500 rpm. Faster CD-ROMs operate at multiples of this speed. So a 4X drive operates at 800 to 4,000 rpm. The strategy used in a CD-ROM is to provide data at a constant rate, which requires the motor to spin slower when reading data at the outer edge of the CD. Since the outermost tracks of the data spiral are much greater in circumference than the inner tracks, they contain more data per spin than the inner tracks do. To keep the data acquisition rate at its highest level, the spin motor speeds up and slows down the spin, depending on where the laser assembly is working. The speed is controlled by sensors on the circuit board holding the spindle motor.

Above the spindle, on the underside of the now-removed metal covering, is a free-spinning spindle to help hold the CD in place. Next to the drive spindle is the assembly that holds the laser and positions it underneath the CD. The assembly rides on two rails.

A small DC motor drives a long worm gear that pulls the assembly back and forth across the underside of the spinning CD. A few screws hold the motor in place. Freed from its perch, the motor still works. I connected

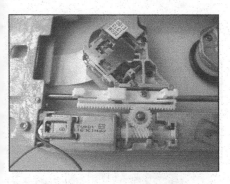

its leads to a 9-volt battery with two alligator clip leads and was rewarded with a whirring sound.

Looking closely at the laser-carriage assembly you can see a lens through which laser light passes, but you can't see the laser itself. It's off to one side and positioned to fire horizontally, not vertically. Digging deeper we will get to the laser, but take a look at a photo of the laser and the lens up close, resting on a dime. They are tiny.

Here, too, is a look at the laser under 20x magnification.

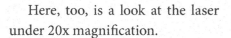

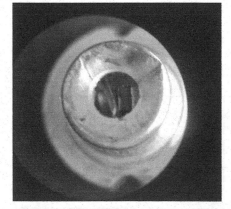

Removing the laser assembly cover allows us to see the laser, an angled mirror, the photo detector, and a lens. The laser and detector are arranged at a 90-degree angle to each other, but light for both devices has to travel through the same lens. Light from the laser (with a 780-nanometer wavelength, the light is just beyond human vision) reflects off the angled mirror in the assembly to a prism that reflects the beam up to the underside of the disk. The data is stored on the underside of the disk.

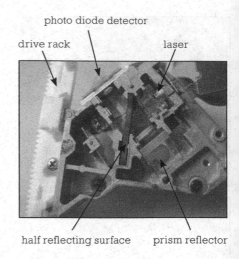

photo diode detector
drive rack   laser
half reflecting surface   prism reflector

Light that reflects off the aluminum coating of the flat surface of a disk (not a pit) passes down through the prism and straight through the angled mirror to the photodiode detector. The system works by detecting the difference in reflectivity between land (reflects 70 percent of the light) and pits (reflects 30 percent of the laser light).

Looking at the front end of the CD drive, you can see the drive belt and pulley that operate the tray holding the CD. The motor drives the tray in and out and also raises the laser assembly from below after the CD is in place. The assembly is mounted on a hinge at the rear, and the front is guided up and down by the angled slot in the black plastic. The piece rising up on the left side tells the circuitry the current position of the tray.

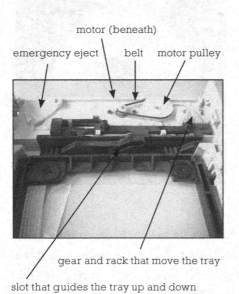

motor (beneath)
emergency eject   belt   motor pulley

gear and rack that move the tray
slot that guides the tray up and down

gear that drives the tray in and out

motor pulley (belt removed)

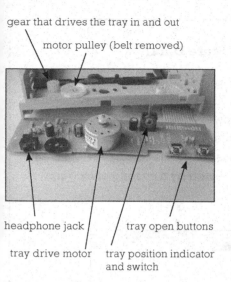

All of this is clearer when you remove the motor from behind the plastic cover.

headphone jack / tray open buttons

tray drive motor   tray position indicator and switch

What Now?

Motors, shafts, pulleys, and gears are useful in many small projects. The metal case could make a nice open cabinet or drawer for holding tools. I've used the laser assembly and ribbon cable to decorate holiday presents—they're much more interesting than standard bows.

Some people use brushless motors in radio-controlled models. However, if you decide to take the motor apart, the magnetic disk from the brushless motor is quite powerful; if nothing else, it will do a nice job of holding your photos to the refrigerator door.

How Do CDs Store Data?

CDs are made of polycarbonate plastic, 120 millimeters across and 1.2 millimeters thick. A laser burns pits in the underside of the CD to record data, and a lower-powered laser bounces a beam of light off the bottom to read the data. The other side of the CD has a metallic layer that reflects the laser light back. The metal layer, usually made of aluminum, is protected by a coating of lacquer.

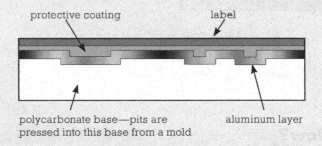

protective coating   label

polycarbonate base—pits are   aluminum layer
pressed into this base from a mold

Pits are about 100 nanometers deep and 500 nanometers wide. A nanometer is 1 billionth of a meter. The data is laid out in a spiral that starts at the inside of the CD and works outward. Separation between two adjacent tracks is about 1,600 nanometers. If you could unroll the spiral path of data on a CD, it would stretch about 3.5 miles.

The pit depth is about one-quarter of the wavelength of the laser that reads the data. Reflected from a pit of this depth the light is reduced in intensity so the circuitry can differentiate between the pit and adjacent nonpitted areas called land. Mastered CDs have the data molded into them from a glass master called a father. Burned CDs carry data stored by a laser that burns a dye layer onto the CD.

The pits or bumps do not represent ones or zeros directly. The change in elevation from a pit to land or land to a pit represents a one, and no change represents a zero.

Incidentally, the first music album released in the CD format was Billy Joel's *52nd Street* in 1982. Three years later, the first computer CD-ROM was brought to market.

CLOCK

Treasure Cache

LCD

9-volt battery terminal

Speaker

Transformer

Tools Required

Flathead screwdriver

Phillips screwdriver

Scissors

Thrift stores and garage sales sell electric clocks all the time. The one I bought for 99 cents still worked—at least the timekeeping part, if not the "soothing natural sounds" feature.

Lefty Loosens

The clock shown ran on either a 9-volt battery or the power from a wall outlet. The door of the battery compartment slid open, but there was no battery
included. There was a battery termi-
nal, though, which I thought could
be useful. I cut the wires, leaving as
much wire as I could.

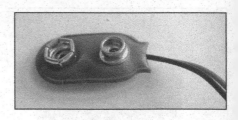

Four screws held the front and
back halves of the clock together.
As soon as they were removed the
snooze alarm button fell away. I bent the LCD screen back to see the circuit
board below. It was held to the back half of the case by a few more screws.

Eventually the entire system of components came out.

The LCD screen that displays the
time connects to the circuit board
by a short, wide ribbon connector.
Scissors cut through the connector.

liquid crystal display

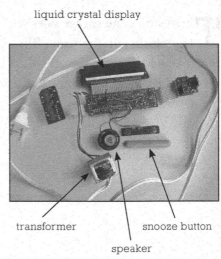

The speaker that plays both the
soothing natural sounds—at least
that's what the outside of the case
calls them—and the alarm is glued
to the back of the case. (I pried it out
with a flathead screwdriver.)

transformer   snooze button

speaker

The buttons you push to set the
time are on a small circuit board
that can be seen on the left side of
the photo. The contact plate is held
in place by three tabs, so it is arched
upward. When you press the but-
ton, you bend the plate down to make contact across the switch on the circuit board. Releasing it allows it to spring back into position.

There is a transformer that con-verts the 110-volt household current into 9 volts. Check to see whether your waste disposal company recy-cles transformers, as it is important to keep these out of the normal waste stream. Older transform-ers, the ones you are likely to find in a morbid clock, contain PCBs, organic chemicals that have toxic effects—clearly something we don't want leaching out of a disposal site into the groundwater.

The main circuit board has a long, integrated circuit that controls the clock and sounds. It is filled with individual components: resistors, capaci-tors, and transistors.

What Now?

Try messing around with the LCD. Using alligator clip leads, connect a 9-volt battery to various pairs of contacts on the LCD to see what lights up. Maybe you can figure out how to make full numbers or letters show up.

The speaker in the clock I took apart wasn't very good. I ripped the clear plastic diaphragm off, removed the electrical connections, and placed it on my refrigerator. It makes a nice magnetic picture frame.

COMPUTER MOUSE (MECHANICAL)

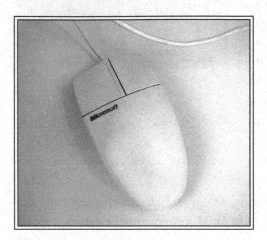

Treasure Cache

   Ball

   Microswitches

   Notched counter wheels

   Plastic case

   Wire

Tools Required

   Phillips screwdriver

These small devices give you hands-direct control of your computer. They make it much easier to navigate where you want to go in a document or web page and to select which of the options open on that page you want.

Lefty Loosens

A single screw held this mouse together. The top cover was clipped to the bottom at one end, so there was only one screw at the other end to hold the two halves together securely. (Other models have two screws.) With the screw removed, the top lifted off.

This mouse has two buttons— left click and right click—and a roller ball. The ball fits into a hole in the base of the mouse; the hole's diameter is slightly smaller than the

microswitches   gates   track ball

light

ball so the ball doesn't fall out the bottom. To keep the ball snugly against the x- and y-axis rollers, a spring-tensioned wheel presses against the ball. This wheel is located at the top right of the ball.

The ball contacts and turns the two rollers. Even with the cover removed you can see how this works by reinserting the ball and gently rolling the mouse across your finger or tabletop. As the ball turns, it spins one, the other, or both of the roller bars, depending on what direction you spin the ball. The roller bars are arranged so that they are perpendicular to each other.

One bar spins only when the mouse is dragged in one direction (parallel to the longer axis of the mouse), and the other spins only when the mouse moves in the perpendicular direction (parallel to the shorter axis or width of the mouse).

The roller bars each have a plastic gear-like device at one end. But unlike gears, they do not mesh with any other gears. Instead, the teeth break a beam of light generated by tiny infrared LEDs. On the opposite

side of the notched wheels are electronic components (infrared photo transistors or detectors) that react to the presence of light. When light shines on them, they pass electricity; when there is no light, they stop passing the current. The pulses that each LED/detector sees are counted and converted through software into movement on the computer screen.

At the top of the ball are two microswitches. These lie directly beneath the left and right click buttons. When you depress the buttons, you activate

microswitches

these momentary contact switches. If you took one of the tiny switches apart you'd see a metal lever making contact to complete the circuit and acting as a spring. As the metal lever moves, it makes the clicking sound you hear when you click your mouse.

To get the circuit board out, the two roller bars have to be removed.

They are held in place by plastic posts. Bending one post away from the roller bar allows you to pick the roller up and out of the mouse. With both roller bars removed, the circuit board lifts out.

All of the electronics are mounted on this circuit board. The chip converts the pulses of electricity it receives from the infrared photo detectors and the switches into code that it sends to the computer.

What Now?

A mouse would make a great switch. With the two microswitches still in place, you could connect them in series to play a chime, turn on a light, power a motor, and much more.

Who Invented the Mouse?

Douglas Engelbart of Stanford Research Institute invented the mouse in 1963. (Before it was called a mouse, the device was "an x-y position indicator for a display system.") Engelbart demonstrated it at a conference in 1968 and was awarded patent number 3,541,541 in 1970.

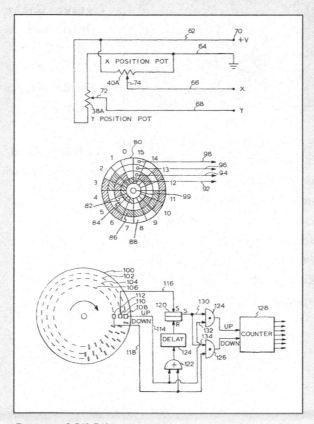

Patent no. 3,541,541

And where did the device get its current name? The small body with its long "tail," the cord connecting it to a computer, reminded engineers at Stanford Research Institute of a mouse.

COMPUTER MOUSE (OPTICAL)

Treasure Cache

Finger wheel

LED (infrared)

LEDs

Notched counter wheel

Plastic body

Wire

Tools Required

Phillips screwdriver

From the outside, an optical mouse looks like a traditional mechanical mouse. Underneath, however, are two openings: one for an LED to shine down onto the desk surface and another for light reflected off the surface to get back into the sensors inside. There is no rubber ball sticking out the bottom as with a mechanical mouse.

Lefty Loosens

The lid on this ingenious device was held by a single screw. Inside, the finger wheel looks similar to the two wheels found in the traditional mouse. The finger wheel shares its shaft with a counter wheel. The notched wheel looks like a gear, but it doesn't mesh with anything. Instead, it separates a tiny LED from a photo detector.

finger wheel   notched counter wheel

photo sensor   LED

infrared LED   CMOS

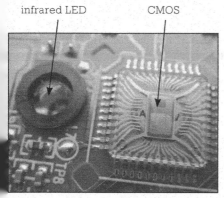

As you roll the finger wheel, it turns the axle holding the notched wheel. And, as that wheel turns, each notch in its perimeter allows light from the LED to reach the photo sensor. The electronic circuit counts the number of times light hits the sensor to move the cursor on the screen the corresponding distance.

Two tiny switches reside directly below the left and right click buttons. When you use the click buttons, you depress the momentary microswitches, completing or interrupting a circuit.

Attached to the plastic circuit board are two LEDs. One faces forward and tells the user that the mouse is receiving power. This LED shines in the visible spectrum. The other LED faces downward and operates in the infrared region of the spectrum.

Removing the single screw that holds the circuit board in place reveals the interesting underside. The infrared LED is provided with a circular lens or cover. Adjacent to the lens is a black square plastic covering. Snip off its plastic legs to find a gold-colored chip underneath. On the right side of this chip is a small square with the photo sensors.

Unlike a digital camera that uses a CCD, a charged-coupled device, an optical mouse uses a different sensor, a complementary metal-oxide semiconductor sensor or CMOS. Both sensor types—CCD and CMOS—convert light into electric charges. Looking through a microscope, you can see the individual pixel elements of this CMOS sensor.

What Now?

The plastic mouse body must have some use, if not for storing things, then for decoration.

DRILL (ELECTRIC)

Treasure Cache

  Chuck

  DC motor

  Gears

  Switch

Tools Required

  Phillips screwdriver

  Rag

Electric drills have largely displaced hand-powered drills and brace-and-bit drills. A basic electric drill is supplied with power by an electric cord you plug into an outlet. These provide lots of turning power and torque—and never get tired—but to use one you have to drag the cord around and be near an outlet. Battery-powered drills are much more popular today. Typically, the batteries fit into the handle. To recharge them, you either plug the drill into a power supply or remove the batteries and insert them into a charging station. With several spare batteries, this latter design keeps you on the job while batteries are charging.

Lefty Loosens

Removing a handful of Phillips screws will separate the two body halves. In the handle are three rechargeable batteries. When they run low, you plug a low-voltage power cord into the drill—these batteries do not come out except if they need to be replaced.

The batteries supply power to the switch, which is directly above them. Pulling the trigger allows power to flow to the motor. The motor is the large silver-colored cylinder above the trigger. Behind the motor is the charging port, and above it is an LED to let you know when the batteries are fully charged.

Below the motor and just in front of the trigger is the reversing switch. Pushed to one side, it sends electricity to the motor to spin it for drilling or screwing. Pushed to the other side, it reverses the flow of electricity, causing the motor to spin in the reverse direction to remove an errant screw or extract the drill bit.

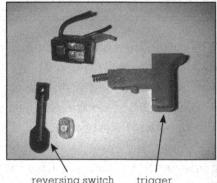

reversing switch   trigger

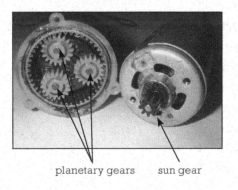

planetary gears   sun gear

In front of the motor—that is, toward the end where the drill bit inserts—is a plastic housing for gears. The motor has a small pinion gear on its shaft that serves as the sun gear in a planetary gear system. The sun gear rotates on the motor shaft and drives the three white gears (the planet gears) inside the housing. The planets all rotate around their axles, which are supported by a disk beneath them, and they also rotate around the inside of the ring.

This arrangement reduces the speed of rotation. The small sun gear, with about half the number of teeth of the planets, drives the larger planet gears with a 50 percent reduction in the speed. The planets rotate around the inside of the ring, further reducing the rotation speed and increasing torque. But wait, there's more!

This is only the first layer (orbit) of the gears. Beneath this galaxy is another identical one. The first layer of planets rotates another sun gear. This sun gear also has three planetary gears that it rotates, which further reduces the speed. The second set of gears provides power for the shaft that spins the chuck holding the drill bits. Grease covers the second set, so keep a rag ready to clean your hands.

This arrangement of two layers of sun and planetary gears appears in electric screwdrivers (see page 173).

The chuck is held onto the shaft by a (left-hand) screw. The threads on this screw were stripped, so I stopped here.

Be sure to dispose of the rechargeable batteries properly.

Who Invented the Electric Drill?

Samuel Duncan Black and Alonzo Galloway Decker did not invent the first electric drill, but their improved design made electric drills much more user-friendly. Their patent (number 1,245,860, issued in 1917) covered the power switch, placing it in the handle as in a pistol grip.

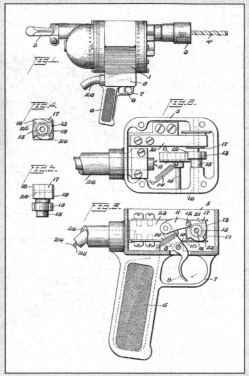

Patent no. 1,245,860

These enterprising machinists started their company, Black & Decker, in Baltimore in 1910. Black had several other inventions, including electric screwdrivers, pumps, reclining seats for touring cars, and the reversing mechanism for electric drills. Decker shared in some of these inventions and had several on his own. These were two creative guys.

What Now?

Test the power switch or trigger. A common failing in electric devices is the power switch. Test the switch with a multimeter to see if pulling the trigger allows current to go through. If the switch is bad, replace it and you've repaired the drill.

The chuck, holding either a drill bit or a tiny flag of your favorite country, would make a nice desk ornament. I use one as a pen holder.

The gears are greasy and would require a good cleaning before reuse.

DUSTBUSTER

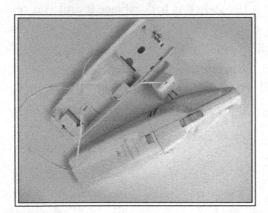

Treasure Cache

DC motor

Fan blade

LED

Rechargeable batteries

Spring

Tools Required

Flathead screwdriver

Phillips screwdriver

Rotary cutting tool

Small hammer

With over 150 million cordless vacuums sold since they were invented in the late 1970s, they show up in thrift stores and trash cans quite frequently. Often you can find an operative vacuum that has been thrown away because the batteries need to be replaced or the charger doesn't make good contact with the battery case inside the vacuum.

Lefty Loosens

The battery case could be exposed without removing all the screws that held the two sides together. With all the screws out, the motor with fan blade, switches, and batteries dropped out. The on/off slide switch fell out along with its return-to-position spring.

The description on the outside of the DustBuster says that it needs 15.6 DC volts to recharge the batteries.

Latches hold the battery case together. A flathead screwdriver inserted between the two halves of the battery case pries it open.

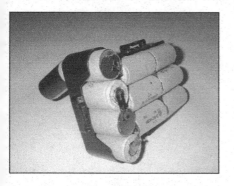

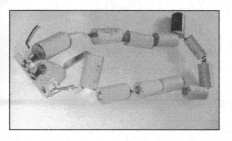

Pulling the batteries out is like extracting sausages from a plastic bag. Out and out they came: 13 C-sized NiCad batteries. Each cell can hold 1.2 volts, so it takes 13 of them to hold 15.6 volts. (Remember to take spent NiCad batteries to hazardous waste collection sites and not let them end up in a landfill.) The batteries in this Dust-Buster work, so they can be pulled apart and used in applications that require rechargeable C cells.

With the flathead screwdriver, the fan blade pries off the motor shaft. The recharging light, an LED, connects to the motor mounting and switch assembly. Its contacts slide off contact tabs on the assembly. Getting the motor off of its plas-

tic mounting and switches is more difficult, but a lot of prying with a small screwdriver and some cutting with a rotary cutting tool will free the motor.

Several very small tabs hold the bottom end of the motor in the metal case. Removing them is a bother; a hammer tapped onto the motor shaft pushes it out the back.

Befitting a motor of this size, the brushes are large. Along with the brushes, the rotor assembly or armature slides out.

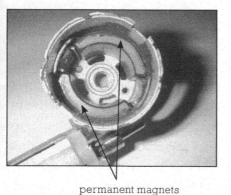

permanent magnets

Two permanent magnets remain glued inside the motor housing.

The motor shaft has fan blades around it to move air and keep the motor cool. At 75 to 80 percent efficient, DC brush motors generate a lot of heat.

Motor Basics

If you're not familiar with DC motors, check out the motor inside a DustBuster. Power is supplied to the rotor through the two brushes. The brushes rub against the commutator (on the rotor shaft), which is a switch broken into several parts. You can see the gaps in the commutator. In this motor there are five separate regions or contacts.

The commutator energizes each of the five electromagnets at just the right time to keep the motor spinning. As it spins, the energized electromagnet is attracted to one of the permanent magnets inside the motor and repulsed by the other. By constantly changing which electromagnet is energized, the motor continues to spin. The motor transforms electric energy into spinning mechanical energy.

What Now?

For a battery-powered motor, this makes a strong fan. If you switched propellers from the impeller that blows air outward (used to draw air into the suction side of the vacuum) to a blowing propeller (available at hobby stores and science supply houses), you would have a great blower. An alternative to purchasing a propeller would be switching the connection to the batteries to run the original centrifugal fan in reverse. Experience suggests it wouldn't do as well as a proper propeller, but it wouldn't hurt to try. Do you barbecue with charcoal? A DustBuster blower would get the burning coals ready for steak in a few minutes.

If you have other projects that require low DC voltage, separate the battery string so you can recharge each battery. Then engineer the battery case to hold the batteries in series to reach the voltage you need. Connect groups of batteries in parallel to increase battery life between recharging.

DVD PLAYER

The DVD player is a marvel of modern technology. If you're not familiar with how DVDs work, you're in for a treat, as opening up a DVD player exposes you to a plethora of gizmos and motors. DVD players show up in thrift stores and in computer recycling centers. Among your circle of techno-savvy friends, someone must have an old DVD player from their computer or entertainment cabinet that is waiting for your screwdriver.

Treasure Cache

   Assorted electronics
      components

   DC motors

   Drive belt and pulley

   Electrical cables

   Fuse

   Gears

   Lenses

   Lever-action switch

   Magnets

   Metal rods

   Rubber gaskets (bushings)

   Self-tapping screws

   Springs

   Switches

Tools Required

   Flathead screwdriver

   Needle-nose pliers

   Phillips screwdriver

   Rotary cutting tool (optional)

   Scissors

   Soldering iron and sucker
      (optional)

Lefty Loosens

A few Phillips screws held the metal case on this player. Removing the case and flipping the mechanism upside down revealed the bottom of a circuit board and two motors.

One motor operates a slide that lifts the assembly up to receive a DVD. A lever-action switch controls this motor. Pushed in one direction, it sends current to spin the motor clockwise; pushed in the other direction, it reverses the rotation. This is a handy switch to have, well worth the time to either cut its circuit board away or to de-solder its leads from the circuit board.

This tray motor comes out of the plastic base with a push or, in some cases, by removing two small screws. This low-voltage DC motor

control switch   tray lift motor

is worth saving. Cut the two thin wire leads as far away from the motor as possible to give yourself as much length as possible for making connections later.

On the shaft of this motor is a small pinion gear that drives a larger gear or a small pulley driving a belt that turns a larger pulley. In either case,

the resulting motion moves a pin left or right. The pin slides through an inclined slot that raises the assembly.

Remove the large circuit board from the bottom of the metal assembly. A few screws hold it in place—not much that is interesting or useful here. A second, smaller circuit board is the power supply. Remove it by taking out a few small screws.

Several of the screws are self-tapping, which makes them more useful to you. Look at the threaded end of the screw. Self-tapping screws are not uniform in diameter toward the end; some taper slightly, and others have a small notch at the very end. These modifications allow you to screw them into a material such as wood or plastic without predrilling the holes.

The power supply has some interesting components. In the lower right is a glass-encased fuse. If something causes a surge in the incoming current, the fuse will burn out, saving the other components downstream. Directly above the fuse is a large, orange rectangular capacitor.

Above the capacitor is an inductor—a coil of copper wire. This, in conjunction with the large capacitor, is used to filter out unwanted oscillations in the current (noise). To the left of the inductor are four diodes. They are the same size and shape as a resistor but have black bodies and silver tops. These form a full-wave rectifier bridge that converts the alternating current from the wall outlet to a fully rectified, but varying, current. So, the current no longer swings positive and negative 60 times a second; now it rises and falls to zero 120 times per second. A very large capacitor (400 volts, 33 microfarads) rises above the plane of the circuit board. Below it is an integrated circuit that controls the operation, along with resistors (cylinders with bands) and capacitors (black cylinders like the large capacitor with small disks on legs).

The large transformer dominates the board. It provides the different voltages required by the various components. For example, the motors require voltages different from what the laser requires.

To the left of the transformer are several capacitors, resistors, and diodes; along with a three-terminal voltage regulator. The smaller integrated circuit is an opto-isolator, a device that converts electricity into light and back into electricity to isolate two different parts of a circuit.

drive motor

The second motor you see from the bottom spins the DVD at speeds between 200 and 500 rpm. In most DVD players it sits directly beneath the DVD so the motor speed is the same as the DVD speed. Two small screws hold the drive motor to the metal frame. To get to the screws you have to remove the spindle (that the DVD rests on) from the motor shaft. Use a flathead screwdriver to pry up the spindle and then unscrew the motor.

The third motor is part of the tracking system that moves the laser assembly or sled. This motor is screwed into the metal frame. It turns a small gear on its shaft that rotates another gear and so on. The last gear in this chain drives a rack, like rack-and-pinion steering in a car. The rack moves the sled along two metal rod rails. Removing four screws frees both rails so you can remove the sled. Pull out the ribbon cable from the socket in the sled.

In the center of the laser assembly is a lens. Light from the laser(s) inside the assembly passes through this lens and reflects off the tiny (400 billionths of a meter long) bumps on the DVD. Laser light that reflects back into the lens is sensed by a photo detector. If the DVD player is also a recorder, light from two different lasers passes through this lens. So, three different devices have to be able to see through this lens: the laser that illuminates the DVD, the photo detector that senses the bumps on the DVD, and a stronger laser that can etch new pits into a recordable DVD. The inside of the laser assembly is going to be an interesting place to explore.

Slide the sled assembly off the rails. Look closely at the lens. Does it move up and down? The lens can be focused. A linear actuator or motor can drive the lens in and out a fraction of an inch to focus it. Copper wire is wrapped around the plastic housing that holds the lens. Passing a current through this wire creates a magnetic field. At either end of the lens are two tiny, shiny rare earth magnets. Just as in a DC motor (see page 47) the induced magnetic field in the copper wires pushes and pulls against the field of the two permanent magnets to move the lens. The magnets are surprisingly powerful; you can pull them out to stick on your refrigerator. The point of a small flathead screwdriver will pry the lens off the assembly.

laser   photo detector

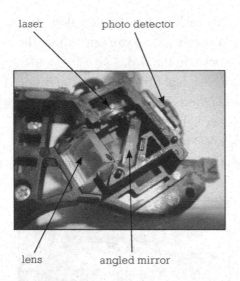

lens   angled mirror

Beneath the lens is a metal body that holds the laser and photo detector. It also holds an angled lens and an angled mirror. The laser is a round component pointing into the center cavity. You can extract the laser by prying on the front to push it out the side of the metal body.

Before you remove the laser, look at the path the laser beam takes. It comes out of the laser that is oriented in the horizontal plane. The beam reflects off the angled glass and then is bent into vertical orientation by the angled lens. From there it passes through the round lens that you extracted earlier.

The photo detector is mounted in the adjacent side of the metal body, on a path perpendicular to the path of the laser light beam. It looks directly at the angled mirror but has to pass through the angled piece of glass (that reflects the laser). By using the angled glass pieces, both the laser and photo detector can "see" the DVD.

photo detector

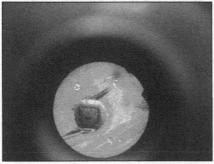

Even with some magnification, you don't see much in the photo detector.

In a DVD player that can write as well as read DVDs, there is a second laser. With the top removed you can see the two lasers on opposite sides of the cavity and the photo detector on the left side looking directly toward the angled lens.

All these components come out with a push or pry: the two lasers and photo detector, with assorted lens and half-mirrored surfaces. In this DVD-ROM, the motor that moves the sled is a small, rectangular motor that drives a worm gear. As the worm gear rotates it moves the sled in and out.

CDs and DVDs: What's the Difference?

The primary difference between a CD and DVD is the wavelength of the light generated by the laser. For a DVD, the wavelength is shorter (650 nanometers) than the wavelength used for a CD (780 nanometers). The shorter wavelength can "see" smaller bits of data, so more data can be squeezed onto a DVD. If you could unwind the track of a DVD, it would stretch seven miles. A double-side, double-layered DVD would stretch four times as long.

What Now?

Worm gears and three motors with gears, belts, and pulleys give you a lot to start with. The motors operate with low-voltage DC power and seem to be robust enough to handle 9 volts, making them easy to connect to power by battery. The rails could be axles for a small project. If the DVD isn't irrevocably disassembled, moving the sled with its motor synced to music or other stimulus would be interesting.

EMERGENCY RADIO

Emergency radios are popular purchases during hurricane season. You can catch the news and weather on an emergency radio using the rechargeable battery or solar cell, or by generating electricity with a hand crank. A few turns of the hand crank stores enough energy in the battery for the radio to play for a few minutes. Even a nonworking emergency radio can be a great garage sale find.

Treasure Cache

Antenna

DC motor/generator

Gears

Rechargeable battery

Solar cell

Speaker

Tools Required

Alligator clip leads

9-volt battery

Phillips screwdriver

Scissors or wire cutters

Voltmeter

Lefty Loosens

Removing four Phillips screws freed the back panel from the rest of the radio.

Inside, the back side of the circuit board dominates the view. The dynamic speaker is the large metallic circle. On the left side is the hand crank. The crank turns a series of gears that are in the plastic housing. At the bottom of the housing is a DC motor. Turning the crank spins the gears that increase the speed of rotation. The output of the gearbox is spinning much faster than you

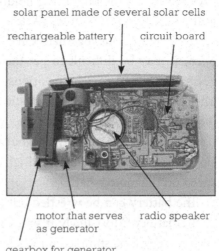

solar panel made of several solar cells

rechargeable battery   circuit board

motor that serves   radio speaker
as generator

gearbox for generator

can turn the crank, and it turns the motor shaft. Spinning the motor shaft generates electricity that the wires carry to the rechargeable battery. Direct current motors, like this one, can either use electricity to spin or use spinning to make electricity.

The hand-crank generator is a nice find. Run a test on it. Clip two alligator clip leads to the motor tabs (where the wires are soldered onto the motor) and touch the other ends of the clip leads to the two terminals of a 9-volt

battery. The crank should rotate, showing that the motor works and that the gear train is intact. Warning: 9 volts is more than the motor/gear system was designed to handle, so don't keep it connected to the battery for too long.

In this model, the hand crank assembly and the motor lifted out. The back cover was the only thing holding them in place. Open the assembly's gear housing with caution. (A few screws held this one together.) Once open, it may be difficult to get the gears back in the right positions. The gears are small, white plastic pieces that mesh together to speed up the rate of spinning from your slow grind on the handle to the high speed on the motor shaft.

The battery can be wrestled out. With a voltmeter, measure the voltage of the battery. Most likely it will be zero or near zero. The battery's voltage rating will be stamped on it. Using either two alligator clip leads or the existing wires, try recharging the battery with either the hand crank or the photovoltaic cells. If the battery holds a charge, use it; otherwise, dispose of it with other rechargeable batteries.

The solar cells slide out from a slot in the top of the radio. Place them in direct sunlight and either measure the output voltage with a meter or try driving the motor by connecting it directly to the solar cell. The solar cell shown worked well.

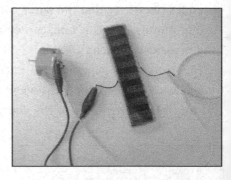

Two screws held the circuit board in place. With these removed, the board can be extracted and the speaker and other electronic components can be salvaged—or tossed.

The collapsible antenna is pretty cool. One screw holds it in place. Cut the single wire that carries the radio signal to the electronics. The antenna is small enough to carry, but can be extended (in this case, reaching 18 inches). Now you have that all-important pointer stick for your next planning session or strategic briefing.

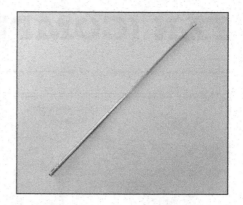

What Now?

The solar cells and motor beg to be used in a model car or boat or a solar kinetic sculpture. The battery and hand-crank generator could have lots of uses, too. Although the hand crank is loaded with plastic gears that you could extract, if it works, it is more valuable as a generator than a pile of gears.

You could also use the solar cell and battery as they were intended: recharge the battery with the solar cell and use the battery to power some other device.

FAN (COMPUTER)

Treasure Cache

   DC motor

   Fan blades

   Magnet

Tools Required

   Flathead screwdriver

F ans don't cool; in fact, they *add* heat to a room. But they do move air, and that can move heat from one place to another. This fan moved heat from inside a computer to outside.

Lefty Loosens

This fan came from a desktop computer. It had no screws or tabs but instead was assembled and glued. I forced the motor assembly off the plastic body and then pried the fan blade off the motor.

Inside the fan is a strong circular magnet. Sitting inside of it are four electromagnets that push and pull against the magnetic field of the permanent circular magnet. That's all there is.

This fan from a laptop was more robust. Instead of a plastic frame, its frame was metallic. It was mounted onto a metallic heat sink and held by a screw. Since metal conducts heat so well, the idea is to draw the heat from the computer to the heat sink and have the fan blow directly on it.

Both fans are brushless DC fans. The rate of spinning is controlled by varying the voltage to the fan. This type of motor is more efficient—meaning it provides more turning power while generating less waste heat—than a normal DC motor.

An Old Invention

Mechanical fans existed before electric motors were invented. The rich had ceiling fans powered by servants. The servants pulled on ropes to move the fan blades back and forth.

What Now?

The fan may be more valuable intact. Try running it with a 9-volt battery. If it works, it will quickly drain the battery, so you would do better to find a low-voltage power supply (wall wart) to power the fan.

You can use a working fan to create an all-important piece of cold-weather gear: a dryer for your winter boots, socks, and gloves. Build a wooden box just large enough to contain the fan at one end. Use a hole cutter to cut an even number of 1½-inch holes in the top of the box, and insert short sections of PVC pipe. Place your wet clothes over the pipes and turn on the fan to dry them. (If a pipe is not in use, push it all the way to the bottom of the box to cut off its airflow and increase airflow to the other pipes.)

Cutting off one or more of the fan blades and then powering the fan provides an eccentric motion similar to the vibrating motor in your cell phone. Attach some springs or stiff wire to four corners to make a jumping kinetic sculpture.

If you do take the fan apart, the permanent magnet may be the most useful component.

FAX

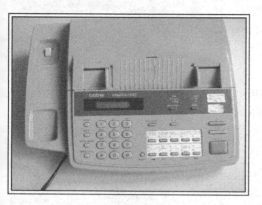

Being able to send documents or images over telephone wires can seem like a miracle. How does it work? Take one of these apart, and you will understand.

LOOK OUT!

The large capacitor on the power supply can deliver a strong electric shock. Be sure to discharge it before removing the power supply. See page 64.

Unscrewed Value Index.. 10

Treasure Cache

Capacitor

Charge-coupled device (CCD)

Fuse

Keys from keyboard

LCD

Lens

Linear actuator

Rollers, rods/axles, springs

Screws

Speaker

Stepper motor

Thermal printer

Wires

Tools Required

Flathead screwdriver

Phillips screwdriver

Scissors

Lefty Loosens

This take-apart will keep you occupied for a while. There are a lot of parts to discover inside. Start by removing the bottom panel. Five Phillips screws held the panel on this unit.

Once inside, most of the components are held in place by plastic tabs. Move the tabs, and the parts come out. First out are two connected circuit boards. All the wires pull out of slots on the two boards, so there is no need to cut wires.

On one side is the speaker that is responsible for those annoying fax sounds. Use the flathead screwdriver to move the plastic tabs apart and pull the speaker up and out. Like nearly all of the other electrical connections, the speaker is connected with a harness that clips into a circuit board.

The power cord leads to a third circuit board, the power supply. Two screws plus tabs hold this in place. Be careful removing this as it has a big honking capacitor. *While holding onto only the wooden, plastic, or rubber handle of a screwdriver,* use the screwdriver's metal shaft to touch the two terminals of

capacitor

the capacitor, the large blue or black cylinder. If sparks fly when you do this, be thankful you took the time to discharge it. The shock wouldn't do lasting harm to you, but it wouldn't feel good either.

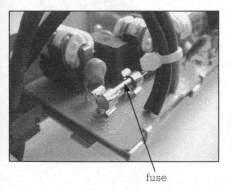
fuse

The power supply board has a glass fuse and a transformer along with a variety of other electronic components.

Also on the underside of the fax are the scanning elements. A fluorescent tube provides the light to illuminate the document being sent. The tube is surrounded by a white reflector and is held in place by a few plastic clips. A wire harness connects it to its power source. Be careful not to twist or bend the tube when you remove it. The tube requires high voltage to operate, and providing this is one of the functions of the power supply removed earlier.

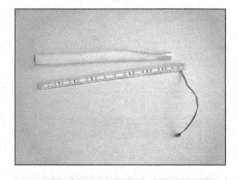

In this model, rollers pull the document to be faxed past the light source. A stepper motor pulls the document step by step through the machine. At each step the light reflected off the document is read by a CCD, or charge-coupled device.

The CCD is attached to a plastic housing screwed into the bottom of the fax. The way to identify this housing is that it contains a circular lens and a reflective glass bar. This mirror is held in place by two metal clips. Slide a flat blade under the clip and pry it up, being careful not to break the mirror. The CCD is mounted on a plastic board that is screwed to the plastic housing. The focusing lens is held in place with glue, and a firm push frees it.

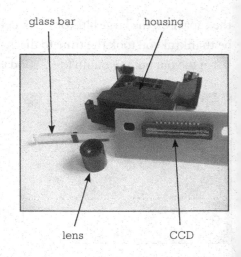

glass bar   housing

lens   CCD

Two groups of wires pass through ferrite chokes to reduce electromagnetic noise. The chokes are not covered, making them easy to see: they are metallic rings. Either pull the wires through them or cut the wires to free the chokes. (See the entry on page 70 for more information on ferrite chokes.)

Turning the fax over to its normal position will allow you to take the keyboard off its hinges. Pry up each side of the keyboard with a flathead screwdriver. On the underside of the keyboard is the plastic film that makes contact when keys are depressed. This might make an interesting coaster for a glass or a gripper for opening slippery jars.

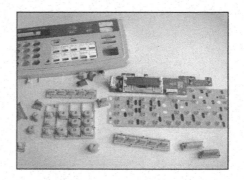

Pulling the film out releases the keys, which fall everywhere.

The keyboard also contains an LCD (liquid crystal display) screen to give the user information on the status of operation. You can remove the LCD by bending back the plastic tabs that hold it in place.

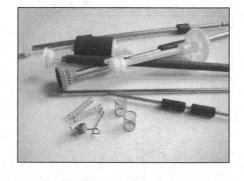

Next the rollers and rods can be removed. This machine moves two different kinds of paper: the documents being scanned and the blank paper that incoming faxes are printed on. Both require powered rollers to move them and require plates and rollers to hold them in place. These all are held in place by tabs at one end and slots at the other. Releasing each one from its restraining tab allows it to be pulled from the slot.

By this point you may be wondering where the printer is. Many fax machines sold today use ink-jet or laser printers, but not this machine. This used the older thermal printing technology. The leading edge from a roll of heat-sensitive paper is pulled past a heating element at the top of the machine to print the incoming fax. Its entire assembly is held in place by plastic tabs and catches. Maneuvering these allows it to come out with the plastic body,

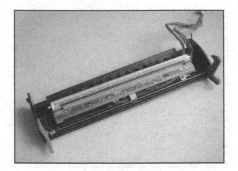

linear actuator   stepper motor

a grounded metal shield, and the heater mounted on a circuit board.

The last major component is the motor assembly—an interesting mechanism. The assembly is held in place on the right side of the machine and comes out intact. A large stepper motor is the prime mover. You can identify the silver cylinder as a stepper motor by counting the number of wires that provide power and control: six. A normal motor would have, in most cases, only two wires powering it, and a servomotor would have three. The motor is held to the assembly only by clips and tabs, which are easily removed.

The motor has to power two separate sets of rollers: one for moving documents being scanned to be faxed and the other for thermal paper being printed for incoming faxes. An engagement/disengagement device can be used to allow one motor to run two different drive rollers; the device itself is powered by a second motor, a linear actuator.

The linear actuator, like nearly everything else in this fax, is held in place by plastic tabs. Prying them out of the way frees this rectangular device. The business end is a plunger that moves outward when an electric current is provided to the other end. The electricity powers an electromagnet that is formed by a cylindrical wrapping of wire.

White plastic gears, all held to the motor assembly by clips, transmit the motor's motion to the two different sets of rollers—one for documents and one for thermal paper.

What Now?

The linear actuator could ring a bell, shoot a marble, or move an action figure. It won't move far but would be easy to control.

The stepper motor has many applications but requires a microprocessor or motor controller to work. The value of stepper motors is their precise movements: each step moves the motor a small fraction of a complete circle.

The lens is a good, but tiny, magnifying glass.

The rollers and shafts would make good high-traction wheels, albeit providing very low clearance.

FERRITE CHOKE

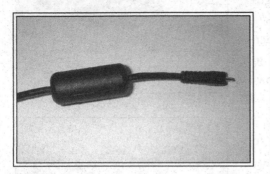

Treasure Cache
  Ferrite choke

Tools Required
  Knife or rotary cutting tool
  Pliers
  Wire cutters

What is a ferrite choke? Without moving, you can probably see one. If there are none in sight, there is probably one hiding only a few feet away. Ferrite chokes are those things at the end of computer and power cords that look like a snake that has swallowed a rat. Maybe you've wondered what those things are—well, read on and you will learn what they do. To add a bit of confusion, these devices are called a variety of names, depending on how they are made. Ferrite toroid, ferrite choke, and ferrite beads are some of the names for devices that perform the same function.

Lefty Loosens

Use wire cutters to cut the plugs or connectors off each end of a wire that contains the choke. Slit open the rubber/plastic cover on the toroid, but *look out: point the knife away from you and anything valuable.* Insert a strong knife blade into the slit to cut or pry the slice open farther. Pushing and prying with the knife is a bit dangerous if your other hand is located downstream. The knife is likely to slip, so beware.

Grab a piece of the covering with the pliers and pull it away. You might want to do some additional cutting to make it easier to pull the cover off.

Insulated wires pass through the toroid. Cut these with wire cutters and pull them through.

The toroid is not a magnet, but magnets stick to it. Inside the toroid? There's nothing; it's hollow.

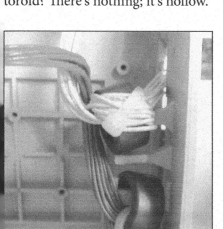

Induction coils or chokes are used in a variety of electrical circuits. Many of the devices shown in this book have small coils inside. Their job is to suppress electromagnetic radiation, and they were first used widely in the early days of television. Later, as computers started operating at faster speeds, electromagnetic interference became a problem, and chokes were added to reduce interference.

The Invention of the Ferrite Choke

John E. Curran and two associates at IBM patented the design for chokes surrounding computer cables in 1970 (patent number 3,516,026).

What Now?

Maybe you could make the chokes into weights for fishing, charms for a geek necklace, or cool-looking wheels for a model car. However you decide to use them, they are interesting to show to people. Most will not know what they are or where they came from. They will be amazed when you tell them that they can find a number of them in their own homes and offices, often in plain sight (albeit covered in plastic).

FLOPPY DRIVE

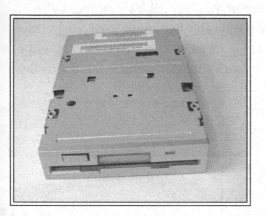

Treasure Cache

   DC motor

   LED

   Metal shafts

   Read/write heads

   Stepper motor

Tools Required

   Phillips screwdriver

   Wire cutters

Floppy drives read and write on floppy disks—or at least they used to. In the late 1980s these drives and the 1.44 megabytes disks they read and wrote on became the industry standard for removable data storage. Today they've been all but replaced by flash drives and CD/DVD drives. That makes them easy to find at garage sales.

Lefty Loosens

This is a great take-apart—I love my job! Removing a few screws allows the protective covers to come off, revealing two motors controlled by a circuit board. One of the motors is a stepper. You can tell it's a stepper by twisting the motor shaft—it doesn't turn smoothly but jumps from one position to the next. The stepper shaft is a long metal rod with a groove cut into it, similar to a worm gear. The advantage of a stepper motor is that it precisely controls the position of the shaft—perfect for finding the microscopic location of the apple pie recipe that you stored on the floppy disk. The motor shaft moves the head assembly inward and outward. The assembly holds one set of read/write heads above and another below the magnetic disk.

The protective cover pops off the heads, revealing two tiny coils of copper wire. These are the electro-magnets that convert electric pulses into magnetic pulses that magnetize the iron oxide on the surface of the disk.

The second motor sits underneath the disk and spins the disk at about 300 rpm. It is not a common type of motor. Its twelve arms, which radiate from the center, are each wrapped in copper wire to make an electromagnet to push or pull the rotor spin around. This type of brushless DC motor is called a 12-pole stator motor. As each magnet is energized, it helps to spin the surrounding permanent ring magnet that is part of the rotor. Rather than being energized through a mechanical contact or brush, this motor is ener-gized electronically. Using a brushless motor reduces energy consumption (as there is no mechanical rubbing by the brushes) and reduces the generation of sparks and spurious magnetic jolts.

LED

In a corner near the motor is an LED, or light-emitting diode. This shines through an opening in the corner of a floppy disk. If light is detected on the other side, the drive will not let you write data to the disk. This is for write protection.

How Close Is the Data?

The heads generate tiny magnetic signals to record data. To realign the iron oxide on the disk, they have to be very close to the surface, within $1/500,000$ of an inch apart.

What Now?

Try messing around with the motor that moves the heads in and out. If you have a microcontroller—a computer from a robot, for example—you might be able to operate the motor. See if you can find wiring directions for a stepper on the microcontroller manufacturer's website or a user's group site.

The magnet from the brushless motor is worth saving. If you don't have an immediate application for it, put it on your refrigerator to hold up your shopping list. Eventually you'll come up with a better use.

GUITAR
(ELECTRONIC TOY)

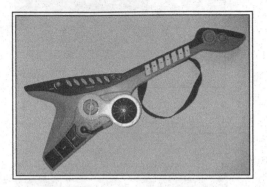

T his instrument has the shape of a guitar, but there are no strings attached.
This is strictly an electronic gadget.

Treasure Cache

Battery case

Electronics components:
    resistors, capacitors, diodes

Knobs

LED

Nylon strap

Potentiometer and switch

Recharging jack

Self-tapping screws

Speaker

Switches

Tools Required

Batteries (4 AA)

Flathead screwdriver

Phillips screwdriver(s)

Rotary cutting tool

Wire cutters

Lefty Loosens

This was a 99-cent find at Goodwill Outlet in Seattle—and it worked! With a change of batteries it came to life. Working or not, apart it came.

First I removed the nylon guitar strap. You never know when you might need one. A phalanx of self-tapping screws held the two halves together. (A large jeweler's screwdriver would have been ideal to reach deep into the screw holes.)

With the back removed, everything is exposed. The speaker is a good-sized 8-ohm speaker useful for many projects. It is held in place by several screws and a plastic retaining ring.

This model has 20 switch buttons visible from the front side. Each makes contact with a rubber dome switch beneath. The dome switches are embedded in a sheet of rubber, and each switch has a carbon center. Depressing the switch button on the guitar depresses a dome switch so its carbon contact pushes against the circuit board. Releasing your finger from the switch allows the rubber to lift the contact off the switch. Pulling the rubber away from the circuit board, you can see how the conductors interweave and how placing the carbon contact on top of them would connect the two sides of the switch.

The larger circuit board inside receives input from all the switches; you can see the conductor traces on the top of the chip. In the center of the board is a raised black dome. This is an epoxy cover laid over a surface-mounted preamplifier. Preamplifiers increase the voltage of a signal, but not the current. An amplifier, electronically downstream of the preamplifier, increases the current to drive the

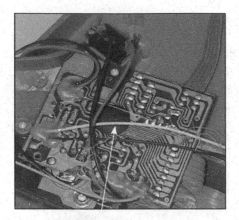

epoxy cover

speaker. Increasingly, circuit board manufacturers are using these epoxy covers to provide protection and heat dissipation for integrated circuits that are expected to last for the life of the device.

The on/off switch and volume control move a 10K potentiometer on the board. The battery case holds 4 AA batteries. Getting this out requires a rotary or other cutting tool. The recharging jack can also be removed; it allows the guitar to connect to AC power through a detachable cord.

The tremolo or trembling effects lever switch is held in by a small screw.

potentiometer knob

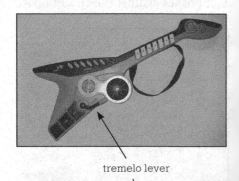

tremelo lever

Circuit Bending

Imagine yourself in a forest with dozens of paths to follow. Each one could lead to a great view or an interesting experience. But you don't know which one will until you give a few of them a try.

In circuit bending, you explore the circuitry of an audio device with a test probe. The probe could be an alligator clip lead or a clip lead biting onto two small screwdrivers, which become probes. Touching one part of a circuit with one clip or screwdriver allows you to make contact with other parts of the device.

There are very few rules in circuit bending, and the only one of life-changing importance is to **never bend a circuit powered by alternating current**. Bend only circuits powered by batteries—direct current. As long as the instrument is powered by a few batteries, the worst thing that can happen is that you fry the batteries or damage some of the circuits of this thing you have already half destroyed. Keep in mind, however, that the sound of a pop, the sight of a flash, the feel of hot batteries, and the smell of burning are all signals to disconnect the batteries.

The second rule is to not experiment with any device that you really should not destroy—like your uncle's favorite electronic accordion. Your goal might not be to break it . . . but things happen.

Circuit bending isn't an entirely random process. Even a small toy or device has dozens of circuits to try, so a methodical search for interesting pairs will be more successful than random touching.

When you find a connection that produces an interesting effect, mark it so you can find it again later. Consider wiring the two points of contact with a switch so you can activate the effect with the push of a button. Or insert a potentiometer to vary the current flow between the two points. A photo resistor could have the same effect, except that the ambient light levels would determine the current flow. Fixed resistors or capacitors between two points may prove interesting.

(continued on next page)

(continued from previous page)

For serious circuit bending, consider having a variety of components on hand—mostly switches. Potentiometers with different resistance ranges, photo resistors, microswitches, and push switches can activate circuits on command. Capacitors, clip leads, and wire allow you to alter the circuitry. Photo cells could be a nice touch to launch a rhythm when the sun comes out; a voltmeter and soldering iron will be needed for that.

What Now?

If the guitar is working, there are many potential uses for sound that comes with the touch of a button. Run wires from the guitar to a new switch you place somewhere obvious—say, on your office door—or less obvious—like beneath a seat cushion on a chair.

Try some circuit bending. Using alligator clip leads, other wires, or your fingers, make connections between two different contact points on the circuit board. Listen for strange combinations of noises you can make. If you find something interesting, use a felt-tip pen to mark the positions of the two contacts.

If the guitar isn't working, try jumping around the on/off switch: clip one end of a set of alligator clip leads to each side of the switch. If the switch is broken, this will bypass it and supply power to turn on the instrument.

HAIR DRYER

Treasure Cache

  DC motor

  Fan blade

  GFCI

  Heating element

  Linear actuator

  Screws

  Springs

  Switches

Tools Required

  Channel locks (pliers)

  Flathead screwdriver (large)

  Phillips screwdriver

  Rotary cutting tool

  Wire cutters

Hair dryers are well represented at thrift stores and garage sales. They hold four components of interest. First is the plug, because it includes a GFIC (ground fault circuit interrupter) switch to interrupt power in case of a short circuit. Second is the motor, which can be quite powerful. Third is the heating element. And last are the switches.

Lefty Loosens

Start with the plug. Bend the prongs outward so no one will come along and exercise their inquisitiveness by inserting this into an outlet. That could be bad.

The plug is surprisingly interesting. Because hair dryers are usually used near water, they operate in a shock-hazard environment. Manufacturers protect you (and themselves, from liability lawsuits) by inserting a circuit inside the plug that will shut off the current if it detects a short circuit. This is a type of GFCI that is common in electrical outlets in kitchens and bathrooms.

This plug had three screws denying access to the inner workings. One was a Phillips screw and the other two were a type of safety screw—in this case, slotted screws with a ridge in the center of the slot preventing a flathead screwdriver from gaining position. Without the corresponding tool I cut the ridge away with a rotary cutting tool to make a slot my screwdriver could get traction in. With that minor detour out of the way, the two screws came out.

With the cover off, a small circuit board and an induction coil at the opposite end are visible. The coil is covered in a hard plastic shell that detects

excessive currents as would occur if there was a short circuit. Removing the circuit board, the coil, and some copper conductors exposes a solenoid or linear actuator.

actuator

The actuator is a simple motor that only goes in and out. When a current is applied to the coils inside the actuator, it creates a magnetic field that drives the piston outward.

actuator arm

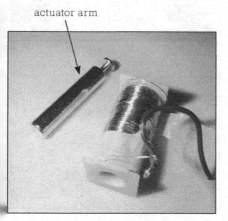

With the metal cover removed and tape unwound, the wrappings of copper wire can be seen. The same type of device is used to ring door chimes.

Now, on to the dryer itself. This model was held together by two screws. Once they were removed the two halves of the handle separated and released the workings. Three switches were in the handle: two substantial switches and one microswitch, all of which could be useful. The switches provide different settings for both the heater and fan.

The plastic housing holds the motor, its cowling, and the heating assembly.

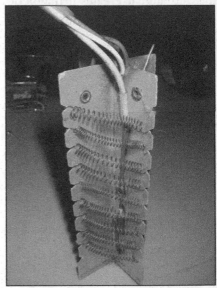

The heating assembly has a heat shield to protect the plastic housing from the hot wires, the wires, and metal frame that holds them. The wires are high-resistance heating wires, similar to those used in electric stoves, waffle irons, and many other devices that produce heat from electricity.

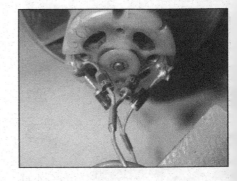

The motor is a DC motor powered by alternating current. At the base of the motor are four diodes forming an H bridge to rectify the alternating current. Although it's normal diet is high voltage, the motor spins when connected to a 9-volt battery. This is a keeper.

The fan blade is securely affixed to the motor shaft. It's secured so well that I destroyed this one in my attempt to remove it.

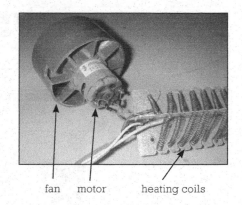

fan   motor   heating coils

What Now?

Need a heater to keep your pizza warm on the way home from the pizzeria? Here are the heating coils you need. The motor/fan could be useful for circulating air, powered by a low-voltage wall wart or battery. I use hair dryers with the heating circuit removed or disabled to power model wind generators using small DC motors and propellers.

HARD DRIVE

Treasure Cache
  Armature
  DC motor
  Magnets
  Platter

Tools Required
  Flathead screwdriver
  Phillips screwdriver
  Torx screwdriver or rotary
    cutting tool

Hard drives are a type of data storage device that retain the data even when the power to the drive is shut off. This is called nonvolatile storage. The drives store information on rigid metallic disks or platters that are spun so read/write heads mounted on an arm can access data anywhere on the platter.

Lefty Loosens

Before opening the hard drive, check out the back side. Here is where connection is made to a power source and where the 40-pin ribbon connection plugs in. The other pins on the back are jumpers, which are connected for specific configurations of the memory.

It isn't easy getting into a hard drive unless you have the proper 6-pointed Torx wrenches. I struggled for quite a while using the wrong tools but eventually got it opened. The easiest way—other than purchasing the tool needed for this job—is to cut slots in the heads of the screws with a rotary cutting tool and then use a flathead screwdriver to remove the screws. But you could go to the hardware store and purchase a Torx driver in the time it would take you to cut the slots.

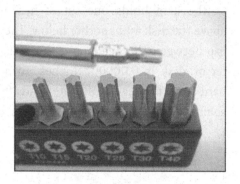

Like other devices the manufacturers don't want you to open, this hard drive has a screw hidden under one of the printed adhesive labels. Push your index finger along the printed paper label until you find a depression. Cut the paper away to access the screw.

Remove the screws to pull off the metal cover. Because the disks are spun so rapidly and the read/write mechanism floats so closely to the spinning disks, the cover is sealed

to keep out dust. On the underside of the drive is the electronics circuit on a printed circuit board and the bottom of the motor that turns the disk.

The top side is much more interesting. Here reside the silver-colored disk and the arm that carries the read and write heads. Disk drives can have several platters or disks, and each has its own set of read/write heads mounted on an arm.

The arm holds one set of heads above the disk and another below. The gap between the heads and the platter or disk is 0.000002 inches. In a

magnet   armature   platter   Torx screw

hard drive the read and write heads float on a cushion of air above the fast-spinning platter. In floppy drives, the heads rub against the disk.

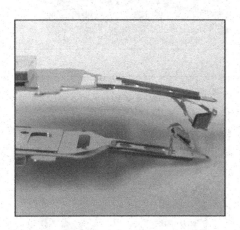

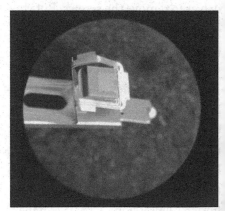

The arm swings on a pivot point without an apparent motor driving it. Removing the arm, one notices the strong magnets in the base of the arm. These are rare earth permanent magnets that form half of the motor. Underneath the arm is a set of coils ("voice coils") that forms the second half. The coils control the arm's movement just like the coils in a speaker vibrate the paper cone to make sounds.

The motor that spins the metal disk spins at 3,600 rpm; however, faster drives can go up to 10,000 rpm. This is a DC motor with fixed-in-place stator windings. A feedback mechanism controls the current to these windings to keep the rotor and the attached platters (which contain the data) spinning at the right speed.

That's Fast!

Spun by the drive motor, called the spindle motor, the disks spin past the head at 170 miles per hour. The disks sit directly atop the motor, so no gears or belts are required to convey the spin to the platters. Spindle motors are brushless DC motors controlled by a feedback system to maintain a steady speed. They are designed to spin for thousands of hours.

The disks or platters are highly polished nonmagnetic metal (usually an aluminum alloy) coated with a material that can be magnetized. The coating is a bit thicker than the space between the head and platter.

In my rush to open up my hard drive I didn't notice the several adhesive patches that covered holes in the metal case. The holes were meant to allow air pressure inside to adjust to atmospheric levels, while the patches filtered air to keep out particles that might interfere with the drive.

What Now?

The platters are so shiny they must have some use. Decorations for your holiday tree? The armature could be a magnetic note holder.

The Invention of the Hard Drive

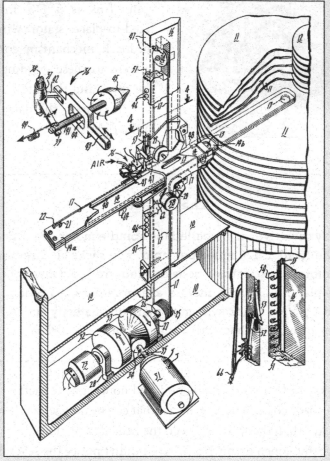

Patent no. 2,994,856

This first hard drive had 50 platters, each 24 inches in diameter. The platters were contained horizontally inside a cabinet that was five feet long, a bit taller than five feet, and wide enough to just fit through a standard doorway. It weighed about one ton. This huge device stored about 4 megabytes of data. Although IBM engineers were sure they could produce drives with larger capacity, no one at the time foresaw a need.

HEADPHONES

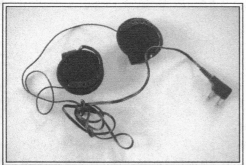

Treasure Cache

Headset

Magnets

Wire

Tools Required

Flathead screwdriver

Scissors

Wire cutters

To transform electric signals into sounds, the signal must pass through a coil of wire surrounding a permanent magnet. This creates a magnetic force in the coil that moves it toward and away from the permanent magnet in sync with the electric signal. The coil is attached to a diaphragm that oscillates in and out as the coil moves, and the diaphragm moves air to generate sound waves that correspond to the electric signal.

Lefty Loosens

I took apart two pairs of headphones. One is the type you get on a long airplane ride and the other is an old stereo headset. Although these are miles apart in cost and 30 years apart in age, they look amazingly similar inside.

The airline headset uses dinky pieces of foam rubber to nestle against your ear. These often fall off the plastic speakers even if you're not trying to take the thing apart. The front and back half of the plastic speaker housing are held together by three clips. I inserted a flathead screwdriver into the slot by one clip to twist them apart.

Wires lead to the speaker, which is covered by clear plastic—the cone or diaphragm—with a central ring made of copper wire in a coil. Cut the plastic away from the outer rim of the speaker with the coil still attached, and all that is left is a magnet inside the center compartment of the back half of the housing. No wonder they give these things away for free.

The stereo headset is not much more complicated. The ear cuffs pull off. A perforated metal plate blocks access to the innards, and this requires popping plastic rivets off. If you slide the screwdriver between the plastic housing and the metal plate, you can use it as a lever to edge the plate up. Inside is a metal frame speaker.

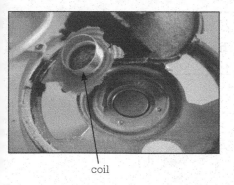

coil

Like its little cousin the airline speaker, this stereo speaker has a diaphragm. This one is made of heavy paper. Cutting the paper away from the metal frame reveals a coil of wire. Directly beneath the coil is a permanent magnet housed in the frame.

The two different headphones have pretty much the same internal components, except that this stereo set also has a small transformer taped to the back of the metal frame. The transformer increases the voltage of the incoming signal to better drive the electromagnet inside. The electromagnet needs higher voltages, but not current, and this is what the transformer provides.

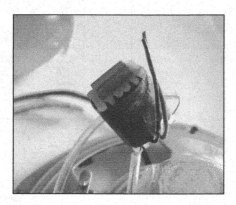

What Now?

If you don't destroy the headset in taking the earphones apart, you might be able to use it to make your own earmuffs. Or, you could create a pair of heated earmuffs powered by electricity! Run some high resistance wire into each cuff and connect it to a battery. Most batteries won't be able to generate lots of heat for very long, but an alternative would be to use a power supply from an old appliance to power the heater—provided you can sit near an electrical outlet outside in the cold. Of course, you probably stand a good chance of burning your ears.

Or, make a cool-looking book stand. Hold three or four of your favorite books upright by clamping the ear cuffs on them. Transform the earphones into tiny speakers by inserting them into plastic funnels from your kitchen.

The Invention of the Loudspeaker

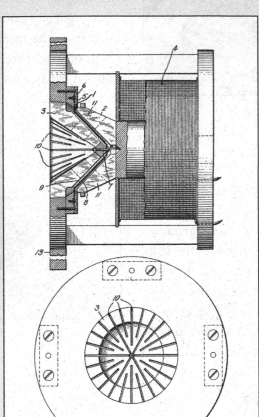

Patent no. 1,631,646

Early radios didn't have amplifiers until Lee De Forest's audion was invented in 1908, so headphones or earphones were used with radios and with telephones. The headphones in use were primitive until the invention of the loudspeaker in 1924. Chester W. Rice invented a design still used today (patent number 1,631,646).

INK-JET PRINTER

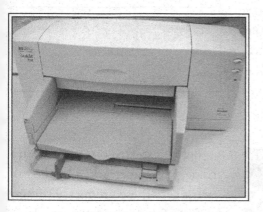

S pouting ink dots smaller than a hair in diameter while flying above a piece of paper at incredible speeds doesn't seem like a good way to print a tax return. But the technology works incredibly well.

Treasure Cache
  DC motor
  Drive belts
  Ferrite chokes
  Gears
  Metal bars/shafts
  Pulleys
  Rubber rollers
  Screws
  Stepper motors

Tools Required
  Flathead screwdriver
  Magnifying lens
  Phillips screwdriver
  Rag
  Scissors
  Torx screwdriver
  Wire cutters

Lefty Loosens

Phillips screws hold the plastic case to the frame inside. Remove as much of the case as you can.

Inside, the electronic circuitry is mounted on a board toward the back of this model. The same board supports the parallel jack for the print cable. The parallel cable connects to a computer.

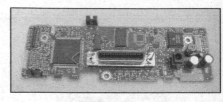

Before I started on the hardware inside, I was sidetracked by the ink cartridges. These are messy things, so if you follow along at home, have a clean-up rag handy. I also recommend doing this outside the visual range of the more fastidious members of your household.

The side of this cartridge could be pried off with a flathead screwdriver. (Other cartridges open at the top.) Inside this cartridge was an aluminum foil bag for the ink—you don't want to puncture this with a screwdriver. I also opened a color cartridge to find three different bags, one for each of the primary colors used in printing.

On the front of the cartridge are the electrical contacts on a plastic film that adheres to the cartridge. You can see them more clearly with a lens or low-powered microscope. They are clearly visible in the 10x magnified photo. These contacts match up to the tiny pins projecting into the cartridge holders inside the printer.

The printhead is at the bottom of the cartridge. Considerable magnification is required to see the openings in the head. In ink-jet printers, the printhead is not mounted on the printer itself, but on the ink cartridge. The printhead is located here because it will eventually clog up with dried ink or dirt and has to be cleaned or replaced periodically, so it is best to dispose of it with the cartridge. This also lowers the price of the printer while increasing the cost of the consumable ink.

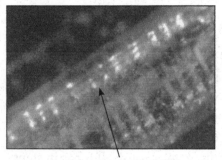

printhead magnified 60 times

To get ink out of the cartridge, a six-microsecond pulse of electricity comes from the printer into the head. The current passes through a tiny three-ohm resistor that heats up enough to boil a microdot of ink out an orifice and onto paper. The departing dot creates a vacuum that pulls the next bit of ink into the resistor chamber, ready to be launched when its time comes. While this printer uses heat to boil the ink and explode it onto the paper, an alternative technology uses the piezoelectric effect: electric current applied to a piezo crystal causes a motion that pushes ink out. Epson uses piezoelectric technology. Ink-jet printers print clearly enough for the naked eye; however, in this image, magnified 60 times, you can see misaligned dots of ink scattered along one side of the two letters.

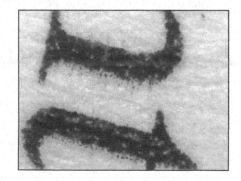

Back to the printer. On the left side are the motor, gears, and control system for pulling paper into the printer. Contact with the paper is made by rubber rollers on long axles that are driven by a stepper motor. The stepper motor drives the axle through a series of gears. This motor is held to the metal frame by two Torx screws. If you don't have a Torx screwdriver, this is your excuse to buy one. The screws come in a variety of sizes, so get a set of drivers with multiple drive heads.

If you're not familiar with stepper motors, check this one out. First, rotate the gear on the motor shaft by twisting the shaft between your thumb and forefinger. You will feel that, unlike motors you're used to, the shaft on this motor doesn't turn continuously. It turns in steps, hence the name. Also note how many electrical wires connect to it. Usually a DC motor will have two wires, but a stepper will have five or

stepper motor

six. Applying power to any pair of the wires will get the motor to move only one step. Operating a stepper requires a microprocessor or motor controller, but the advantage of this type of motor is the precision of its movement.

The stepper motor has a small gear on its shaft, and this meshes with the first gear in a train. The final gear is co-mounted on a shaft with a clear circle of plastic with regularly spaced lines around the outer perimeter. A magnifying lens again helps to make these visible. The plastic disk rotates through a black device mounted above the motor. A single Torx screw holds this to the frame. Inside is an infrared LED and a photo sensor. As the plastic disk

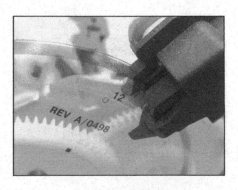

rotates through the photo sensor, the sensor detects infrared light until the beam is broken by one of the tiny lines printed on the disk. The circuitry counts the number of times the light is interrupted to know how far the disk has turned and how far the paper being loaded into the printer has traveled.

The gears are held onto the axle by friction alone. Pull them off and remove the axle as well.

Next, let's look at the ink cartridge carriage. The motor that moves the carriage sits on the left side at the back of the printer. It faces toward the front of the printer and has a pulley on its shaft. It is held to the metal carriage frame by two Torx screws. Disconnect it from its electricity supply and note that unlike the stepper motor, this one—a DC motor—has only two wires.

With the motor removed, the carriage frame can come out along with the carriage where the ink cartridges sit, the metal shaft the carriage rides on, a ribbon cable that carries electric signals to the ink cartridge, the drive belt for the motor, and a clear plastic strip.

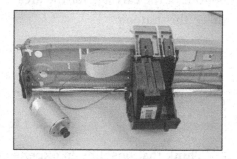

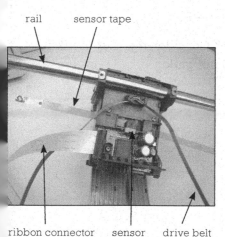

rail   sensor tape

ribbon connector   sensor   drive belt

Slide the metal shaft out of the frame to free the carriage. Cut the ribbon cable with scissors. Open up the carriage, which is held together by Torx screws. Inside of the carriage is a sensor that reads or counts the fine lines printed on the plastic strip. Pry this out and take a close look. With a fine blade, pry the sensor open. Like the sensor on the paper advance mechanism, this sensor helps the printer know where the carriage is, left to right on the page. It too uses infrared light to count how many times the beam of light is interrupted by the black lines on the plastic strip.

On one side of the carriage is a stepper motor, which comes off with the removal of two Phillips screws. Underneath the motor is a series of gears that park the head(s) into rubber gaskets when the printer is not being used. The gaskets keep air out so the ink in the head won't dry out and clog the orifices. The gaskets are messy, so I suggest pitching them. The gears, however, are worth prying off.

The carriage not only delivers the ink-jet printhead to where it is needed, it also collects errant and dried ink from the cartridge. Look at—but don't touch (unless you want inky fingers)—the back of the bottom of the carriage for the area called the spittoon.

Gas Is Cheap

Think that gasoline is expensive? Consider the cost of ink for ink-jets. It costs more than $6,000 per gallon.

What Now?

The stepper motors are useful in projects requiring precision motion. Many robots, for example, use steppers. They do require microprocessors to drive them, but you can find instructions on the Internet on how to build your own controller. With the drive belts and steppers and DC motor, you're ready for a serious kinetic project. Keep the screws that held the motors in place in the metal frame. The plastic control strips are a very tough material, and with some additional thought, you'll likely be able to put them to good use.

JOYSTICK

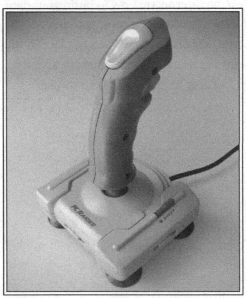

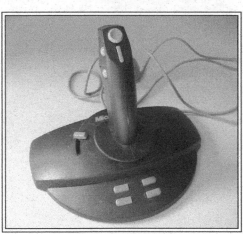

Treasure Cache

   Microswitches

   Potentiometers

   Springs

   Timing chip (555)

Tools Required

   Phillips screwdriver

   Wire cutters

Used mostly for games, joysticks are a lot like mice: they're another way to reposition the cursor on a computer screen.

Lefty Loosens

I took apart two joysticks, both of which are shown here. In the first one, three screws held the handle together.

Inside the joystick are two switches: one operated by the thumb on top of the handle and the other pushed by the forefinger. Depressing these buttons rotates the plastic part inward to push down on a spring inside. When it is depressed, the spring makes a pleasant "sprong" sound as it closes the switch. Releasing pressure on the switch allows the spring to rebound and the switch to open.

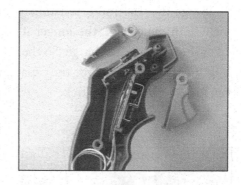

The base is held on with four screws; when they are removed we can see how the gimbals work. The joystick can pivot in either or both of two perpendicular directions. The handle attaches to a ball that rides in a socket and is free to turn. As it turns, it pivots a clear piece of plastic that crosses the bottom of the handle. Pushed in the perpendicular direction, the bottom of the handle slides along the clear plastic piece without turning it and pivots the other (perpendicular) side. Each side connects to an arm that also pivots.

ridge slides between the gap in the clear plastic in one direction and catches the plastic in the other direction

bottom of handle   timing chip (555)

variable resistors or potentiometers, one for each direction

So what does all this movement get you? As one side pivots, it moves a slide along a resistor. Moved in one direction, the slide gives the circuit more resistance, and moved in the other direction, less. I measured the resistance of the material at its maximum width and found about 80,000 ohms. Thus, the movement of the joystick determines the resistance in two electrical circuits from 0 to 80,000 ohms.

The chip in the circuit board is a simple timing chip (called a 555). It sends out signals and measures how long it takes for the capacitor to charge up. The more resistance there is in the circuit, the longer it takes the capacitor to charge. So the

length of time required for the capacitor to charge is established by how far you have pushed the joystick.

buttons   spring

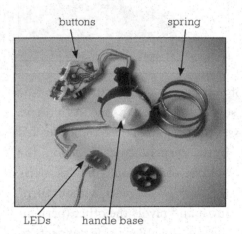

LEDs   handle base

Now switch to a different joystick. The Microsoft optical joystick has two LEDs mounted on the bottom of the handle. Below them and fixed in place is a photo sensor that picks up their movement. A third LED can be pivoted from the side, presumably to calibrate or adjust the sensors. A spring maintains upward pressure on the handle.

The handle carries several buttons, beneath which are momentary contact switches. The thumb button can pivot in two directions. Underneath the button are four momentary contact switches to detect the direction in which you are pushing the button.

contact switches

Types of Joysticks

The first joystick described above is a resistor-capacitor joystick, similar to the original "television gaming apparatus" invented by Ralph Baer, William Rusch, and William Harrison in the late 1960s (patent number 3,659,285). As you turn the knob, you move one of the potentiometers that change the resistance in the circuit and reduce the current, which increases the time required to charge the capacitor. The circuit counts the time delay to charge the capacitor and uses this information to figure out where the cursor should move.

Not all joysticks use this resistor-capacitor method. Some use piezoelectric sensors, and others, like the second device shown in this section, use LEDs and photodiodes. The piezoelectric sensors are made of a crystal material that converts mechanical pressure to electric current. They are also used in electronic devices to make those annoying beeps: they vibrate air molecules to make sounds when an electric signal is applied to them.

Another type of joystick uses LEDs. Pushing the handle interrupts the beam of infrared light detected by a diode or transistor. The circuit counts the number of interruptions to determine where it should place the cursor.

Force-feedback joysticks respond to movements in the game by moving the joystick. The game sends a signal to activate one of a limited number of joystick movements. The movements are generated by two motors, one in each axis. Thus as a building crumbles around you in a game, for example, you can feel the vibrations building up to a large thud. Or, as you push your video game rocket ship to move in one direction, some dark force could resist the movement of the joystick.

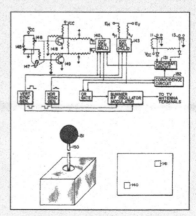

Patent no. 3,659,285

What Now?

The 555 timing chip has dozens of uses. It can generate tones, turn LEDs on and off, and perform many other functions. The variable resistors can control lights, sound level, or motor speed; they waste electric energy, but if your application isn't going to run very long, the loss is not important. The switches are useful in any project requiring control of circuits.

KEYBOARD

Treasure Cache

Keys

Rubber pad

Tools Required

Flathead screwdriver

Phillips screwdriver

K eyboards are banks of 101 or 104 switches that allow users to enter electronic signals into a computer or other digital device. The first keyboards, on typewriters (see page 192), were mechanical systems of levers and springs. Computer keyboards continue to evolve into simpler designs using less expensive materials.

Lefty Loosens

Get your wrist warmed up, because you'll have to do some wrestling with screws. One of the keyboards I disassembled had over a dozen Phillips screws holding the upper and lower parts together. Two of the screws were hidden beneath paper labels glued to the bottom of the keyboard. If you

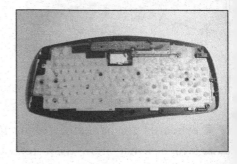

have removed all the visible screws and still the two halves do not separate, remove any paper labels or use the sharp point of a flathead screwdriver to search for screw holes beneath paper labels. Another favored spot for hiding screws is beneath the rubber feet or pads on the bottom of devices.

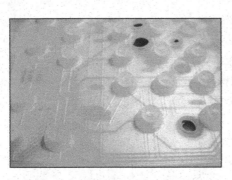

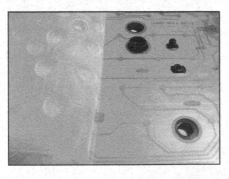

This keyboard has rubber dome switches. A sheet of rubber covers the contacts, and the dome of rubber lying underneath each key restores the key after it has been released. Beneath the layer of rubber lie two contact sheets of plastic film. Each sheet provides one side of each switch that is activated by depressing a key. These pull out of the chassis.

The only other part is a circuit board that has a microprocessor that interprets the signals from the switches and translates them into a code it sends to the computer. In most keyboards the signals are relayed to the computer by a wire that plugs into the back of the computer. In this model the signal is relayed wirelessly using radio waves.

In older keyboards the keys are more complex and use more materials. Some have a metal spring under each key. This model has interlocking plastic legs that act as a spring to return the key to its normal position. When a key is depressed, the two sides of the corresponding switch make contact.

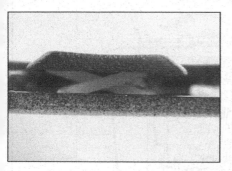 

What Now?

The keys are fun to have so you can write notes on your refrigerator. Glue squares of refrigerator magnet to the backs of the desired keys. Some keys have stalks protruding below the sides, and you might have to cut these off so the keys sit well on a flat surface.

The sheet of rubber could make a gripping pad for jars that are hard to open. Or it could serve as a no-skid surface for plates on a table.

A Pattern of Letters

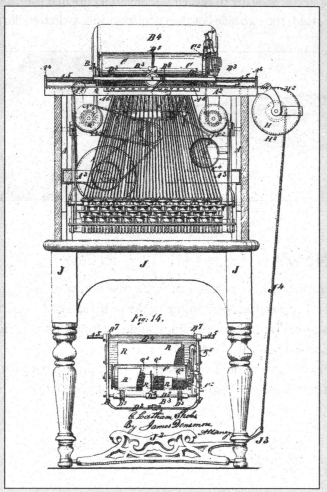

Patent no. 207,559

Christopher Latham Sholes invented the first successful type-writer, from which keyboards later evolved. Sholes came up with the QWERTY format for the key layout, which we still use almost a century after he designed it. (Why is it called QWERTY? Look at the top row of letters on your keyboard.)

LAPTOP

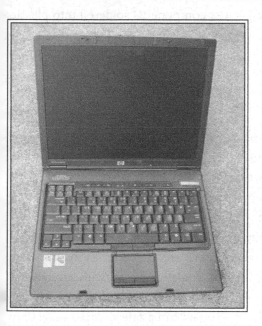

Treasure Cache
    CD/DVD player/drive
    Floppy drive
    LCD
    Rechargeable battery
    Touchpad

Tools Required
    Flathead screwdriver
    Phillips screwdriver
    Wire cutters

This is (or was) an HP Compaq Business Notebook with an INTEL Pentium M 750 processor with 1.86 gigahertz speed. It was quite a machine in its day, not so long ago. The display is an ATI Mobility Radeon X00 TFT active matrix LCD screen.

LOOK OUT!

The tiny fluorescent bulb that lights up the screen is easy to break, leaving ragged edges that can break off in your finger. (The display screen is covered in detail on page 115.) If you open the battery, avoid opening the individual cells, and make sure you keep them out of the normal waste stream.

Lefty Loosens

Circuit boards are stacked on top of circuit boards and are crammed in the case with a DVD drive, expansion port, modem, and more. Most of the circuit boards aren't interesting—just components soldered into plastic boards—and not easy to identify.

One way to figure out what does what is to do an Internet search for the part number or company name shown on the chips.

I removed the battery case and opened it up. Inside are lithium ion batteries, which are known to explode if overheated or over-charged. They also present a prob-lem in disposal: you pay to buy them, and you pay to get rid of them. There are six batteries in this case, each pair being wired in parallel. There is also a board with electronics components inside the battery case. The circuits keep the batteries operat-

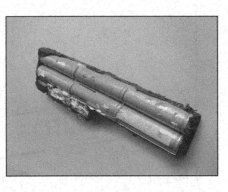

ing within a safe voltage range: they shut the laptop down if the charge drops below 3.0 volts and if they are charged higher than 4.2 volts.

Even with the limited voltage range and the problem of battery disposal, lithium-ion batteries are favored because they supply power for a longer time than other batteries and they don't have the problem (that NiCad batteries suffer from) of memory effect. With NiCad batteries, unless you completely discharge the batteries before recharging them, their maxi-mum power decreases with each charge. This doesn't happen with lithium-ion, which should be kept charged and not allowed to fully discharge.

Removing a few screws and a little prying with a screwdriver separates the two parts of the base of this laptop. Once the keyboard comes out, the power switch is exposed. The switch you push moves up and down to activate a microswitch beneath. This is a very simple and inexpensive solution to the design program of turning on electric power. (This microswitch was too small for me to use, so I didn't remove it.)

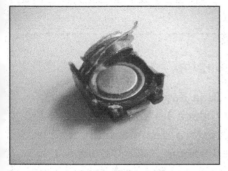

This laptop's tiny, dynamic speaker makes use of a coil of wire wrapped around a magnet. The wire pushes on a tiny, clear piece of plastic to vibrate it at the frequency of the sounds.

On the left side are a fan and heat exchanger. The fan draws air up from beneath the laptop, and blows it through the fins of a heat exchanger, which looks like a tiny radiator and is attached to a curved piece of metal copper tubing that sits directly over the CPU. Inside the tubing are several small wires, presumably to also carry heat away from the CPU. As the CPU warms up, the heat conducts through the metal to the fins where it is removed by the stream of air generated by the fan. CPUs use enough electricity and generate so much heat that they require cooling. This is especially critical in the tightly packed interior of a laptop.

Not all laptops have coolers like this. Many just have a fan to draw air from outside and pass it over the components.

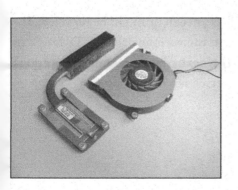

Laptops have a secondary battery that provides power to keep track of time and to power the computer settings. In this computer the clock battery is a 3-volt lithium (not lithium-ion) battery. These batteries are needed in computers that use volatile memory (RAM) for storing the system configuration in BIOS (basic input-output system).

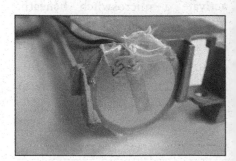

To learn about the touchpad, see the separate entry on page 188.

The First Laptop

Laptops were first suggested by Alan Kay in 1968. Four years later, he fleshed out his concept, which he called Dynabook. Several companies worked to create a commercially viable model. Osborne computers stated selling their Osborne 1 in 1981. Heavy (over 20 pounds), with a tiny five-inch screen and no battery, it was a far cry from what we expect today in a laptop. But it was portable, and it worked. Epson, HP, and Radio Shack soon released their own versions.

What Now?

The circuit boards can go directly in the trash can or might be recycled. The most interesting components are those in the various drives (CD, DVD, or floppy) and in the laptop screen (see the next page). The fan, if you don't destroy it, could be used either as an external fan or as a motor.

LAPTOP SCREEN

Treasure Cache

Cold cathode fluorescent
lamp and inverter

Glass plate

LCD

Polarizing filters

Tools Required

Flathead screwdriver

Phillips screwdriver

Wire cutters or scissors

The laptop from the previous entry had, until I ripped it apart, a TFT active matrix LCD screen. The TFT active matrix is a technology for improving the quality of the LCD (liquid crystal display). TFT stands for "thin film transistors." There is one transistor for each pixel element in the screen. The transistors are created out of a silicon film that is applied to a glass panel. When voltage is applied to one row and one column in the screen, the transistor at the intersection switches on and charges its companion capacitor. The capacitor provides the charge to the liquid crystal, which causes the crystal structure to change, allowing light to pass through. Of course you want color, so instead of one transistor, capacitor, and liquid crystal element for each pixel, you need three of each to represent red, green, and blue. To support full resolution on this screen, there are well over 2 million transistors.

LOOK OUT!

Be careful of the cold cathode fluorescent lamp tube inside the laptop screen. First, be sure not to break the glass and cut yourself. The tube is very fragile and breaks easily. Second, treat the bulb like any other mercury-filled fluorescent tube and dispose of it accordingly.

Lefty Loosens

I started unscrewing my laptop by focusing on the bottom half—the actual computer—leaving the screen for later. However, nearly all the electronics to support the screen are in the bottom. Most notable is the video card. This is mounted above the motherboard, along the back, nearest to the screen. It has a sheath of wires connecting it to the screen. The video card takes the digital image and converts it to signals that the monitor can display. It has its own processor and digital-to-analog converter. The processor takes the computational load off the CPU so it can attend to other work.

A screen with four million pixels (1280-by-1024 resolution; typical for a 17-inch screen) and three colors has 12 million elements to supply 60 times each second. That's over 700 million calculations each second. And this is not simple math; these calculations require multiple operations.

The other board is the display inverter. This is a long and narrow board that sits along the back of the bottom half of the computer. This board supplies power to high-voltage-loving illumination bulbs in the back of the screen. It takes the low voltage from the batteries, converts (inverts) it to alternating current, and raises the voltage high enough to energize the bulbs.

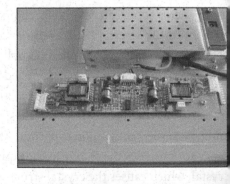

Attacking the screen, we first remove it from the lower half. A few screws hold it to the hinge assemblies. The front glass plate, or mask, comes off revealing a multilayered sandwich of components. These

include plastic film (polarizing films), panes of plastic, a piece of white paper, and the liquid crystal screen itself. One of the panes of plastic diffuses light, spreading it around more evenly.

Most interesting is the screen. Magnified 60 times, we can see the individual pixel elements for each color.

At the bottom of the panel is a long, very thin fluorescent light-bulb, called the back light, in a reflective metal trough. It is a cold cathode fluorescent lamp. Inside is mercury vapor that is energized by the high voltage of the inverter. Once raised to higher energy levels, the mercury gives off invisible ultraviolet light. Invisible light isn't too helpful when you're trying to read the tiny print on the screen, but the invisible light passes through the coating on the inside of the bulb. The coating is energized by the ultraviolet light and gives off light in the visible spectrum—that is, it fluoresces.

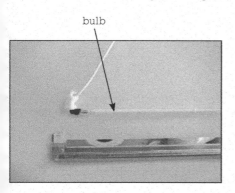

bulb

The bulb is housed in a highly reflected metal tray that sends the light up and throughout the back of the screen. A white background reflects the light forward onto a polarizing filter, which aligns the light.

Dispose of the bulb with a thought as to where you want the mercury inside it to end up—probably not in a landfill near you.

What Now?

The polarizing filters are neat to mess around with. Hold two of them back to back and rotate one while looking through them. See how they block light at certain rotational angles.

The Invention of the Liquid Crystal Display

An Austrian botanist and chemist, Friedrich Richard Reinitzer, first discovered liquid crystals in 1888. Others made advances in the field, but not until a century after their initial discovery did liquid crystals come into their own. In the United States, James Fergason (U.S. patent number 3,731,986, filed in 1971) discovered the twisted nematic field effect used in modern LCD screens. At the same time, a Swiss team filed a patent for the same phenomenon. Fergason built the first LCD based on this technology, and it was used in watches. He was a prolific inventor with over 150 U.S. patents to his credit. In 1998 he was inducted into the National Inventors Hall of Fame. (This author was the founding director of the NIHF.)

LASER PRINTER (LED BAR)

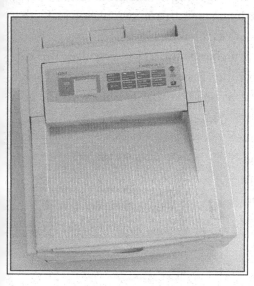

Treasure Cache

    Fan

    Ferrite chokes

    Fuses

    Gears

    Metal bars and rods

    Microswitch

    Rollers

    Springs

    Stepper motors

    Switches

Tools Required

    Flathead screwdriver

    Phillips screwdriver

    Wire cutters

It seems such a short time ago that technology writers were yelling about lasers being an invention without an application. Now they're everywhere. Many laser printers have one laser that is fixed in position and a rotating mirror that sends the laser beam to the exact spot where each period goes (see the next entry, page 125). But this less expensive model uses a bar filled with tiny LEDs instead.

Lefty Loosens

Much of this machine comes out without turning a screw. First up are the paper trays, which lift out. One interesting thing here is the adjustable paper guides. A single gear, held in place by a screw, turns two plastic racks that move in and out so you can use paper of different widths.

The top lifts up so the operator can change the toner cartridge. This cartridge is a cylinder with a center opening. The toner falls out into the trough below. The plastic trough has a gear at one end that meshes with the gear on the cartridge to rotate it. This moves the toner around inside so it can fall out of the cylinder. I would have liked to open the cartridge but the thought of cleaning up all the toner dissuaded me. I carefully put the cartridge in a plastic bag.

If you are following along at home, take your printer apart someplace where clean-up will be easy. The toner—a mixture of dye and plastic—creates quite a mess.

I unscrewed the plastic trough and pulled it out. The lid comes off, and underneath is a mound of toner, ready to go everywhere. Behind the toner is the cylinder (the photo-sensitive drum) where the writing on paper occurs. Above it is a long roller. Both items would be great to see up close, but I opt out and put the assembly in the plastic bag. I'll get a closer look at the second laser printer in the next entry (see page 126).

roller   drum   toner in trough

Directly behind the toner trough is a black plastic assembly with a yellow caution label. The label says that the part, the fuser, is hot. Undoubtedly you've felt printed paper coming out of a laser printer, and you know that it can be warm. Here is where it gets hot. Two screws and clips hold the piece in and with them removed, out comes

the fuser. The fuser is a long metal cylinder, coated with Teflon, that rotates against the paper that has just picked up toner from the drum.

Inside the cylinder is the heating element, a sealed quartz tube with titanium coiled wire inside it. (Since the manufacturer was identified on the tube, I looked it up on the web to find that the little bump in the middle of the tube is where it is filled with gas, specifically, nitrogen, argon, and halogen.) The conductors are made of molyb-

denum, and the titanium is doped (impurities are added to change the electrical properties). It is a very hot incandescent light whose only purpose is to heat the surrounding metal tube so the toner melts onto the paper.

Midway along the length of the fuser is a silver-capped black cylinder. This is a thermostat that controls the temperature of the fuser. (I took it out and forced off the silver cap, which wasn't easy.) Inside is a bimetallic disk that flexes when it is hot. As it flexes, it pushes a plastic piece down to break contact below. Looking inside the thermostat, you can see the contacts. This same device is used in many kitchen appliances such as waffle irons and coffeemakers.

So where does the laser come in? It is mounted on the underside of the flap I raised to get to the toner and fuser. The laser is mounted in a metal case that is held in place by plastic clips. Get-

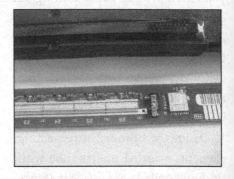

ting the metal case out is easy, but getting the laser out requires some prying and pushing. Eventually the plastic bar comes out, and in it is an array of LEDs (light-emitting diodes), not really a laser.

In this machine there isn't one laser that does the writing. Instead there are many LEDs mounted on the long bar that do. There can be as many as 600 LEDs per inch. The plastic piece that holds the bar has many tiny lenses that direct the light from the LEDs to their proper place on the page.

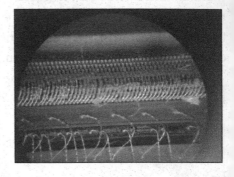

The LEDs shine lights for the dots that make up an image or word onto the drum (discarded above), and this causes toner to be held to the drum by static electrical attraction. Paper rolls over the drum and pulls the toner off the drum by even stronger static electrical attraction. The plastic in the toner melts as the paper moves under the fuser and roller. It doesn't

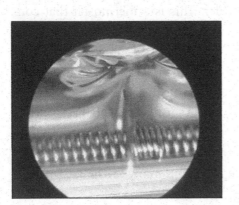

sound to me like it should work, but out comes the printed page.

I also looked at the quartz heating bulb under a low-powered microscope. Very cool detail.

The top of the machine has input buttons and an LCD. The underlying circuit board has two large integrated circuits and a few push buttons plus the LCD.

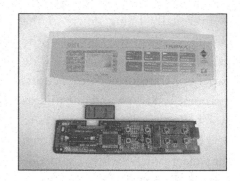

Taking off the sides by removing one screw on each side and then lifting and pulling toward the front reveals two stepper motors. Each is held in place by two screws. The top one is accessible, but the bottom one is blocked by part of the metal frame.

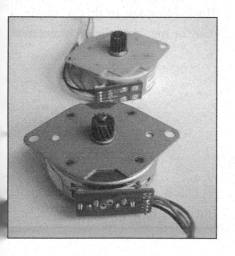

Unscrew the metal frame to collect the steppers.

Eventually I dug down to the circuit boards at the bottom of the printer. They are loaded with a wide variety of transformers, capacitors, resistors, and fuses, and one micro-switch. The microswitch appears to be a safety switch that shuts down the electronics when someone exposes the board.

I've neglected to mention the several bars with rollers and springs galore. This printer provides a major haul of components.

What Now?

The stepper motors are the most valuable components found in a laser printer. They would probably cost $25 or more each. Use them in a robot, automaton, or other gizmo where control of motion is important. Rollers and bars make good wheels and axles. The cooling fan in this machine is rated for 38 volts, so it didn't have much use for me, but other printers may use lower-voltage fans.

LASER PRINTER (SINGLE LASER)

Treasure Cache

DC motors

Drive belts

Electrostatic drum

Fan

Ferrite chokes

Gears

Heating drum

LCD

Lenses

Metal struts and bars

Microswitches

Mirrors

Photo detector

Rollers and bars

Solenoids

Springs

Tools Required

Flathead screwdriver

Phillips screwdriver

Pliers

Rotary cutting tool

Torx wrench

Wire cutters

After taking apart the first laser printer—one that used diodes and not a laser to charge the drum—I had to look inside a printer that had a laser and rotating mirror. Taking apart the second printer was a lot of work, and made quite a pile of useless parts to throw away, but it was a great hike through a very dense technology landscape. Many of the mechanisms are similar to the previous laser printer, so I won't repeat them. What follows are principal differences.

Lefty Loosens

Before taking apart your laser printer, lug it to an area that will be easy to clean. Once you get inside, you will find that toner tends to fly everywhere, and that's not something you want on your carpet.

This machine I disassembled was a load, big and heavy. Unlike its lighter cousin (see previous entry), the sides did not unclip—everything was held in place by screws. I started by taking off the bottom panel, which exposed circuit boards and little else.

I flipped it back so it was right-side up and opened up the top. A finger-operated lever let the top swing up. Inside was the heater and the laser assembly. Both were marked with warning labels, but with the power off and power cord removed, there was no danger of being burned or getting zapped by the laser. The laser assembly was covered with a plastic lid that was held down with a security screw. I didn't have a tool that fit this screw, so I popped the lid open by leveraging it with a flathead screwdriver. The alternative is to cut a flat notch in the head (with a rotary cutting tool) and use a flathead screwdriver to back out the screw.

laser assembly   electrostatic drum

A metal plate, screwed to the plastic body beneath, keeps two lenses in place. With the plate removed, you can see the laser in a brass housing on one side. It shoots down an opening in the plastic body and bounces off a mirror that is spun by a motor below. The laser beam goes out through two lenses and reflects off a long mirror and through a narrow panel of glass known as the light gate.

cover for rotating mirror   lens   laser

The mirror has six sides. It is held onto the motor shaft by a small clip. (I used a flathead screwdriver to break the clip and release the mirror.)

Extracting the laser may require cutting slots in the safety screws in the brass enclosure, as I did. A lens is screwed into the brass body and comes out with a few twists.

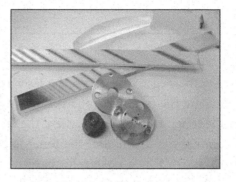

drive gears

I removed the drive gears that turned the electrostatic drum, the yellow cylinder beneath the laser assembly. The motor that drives these is quite large and very well anchored to the frame of the printer.

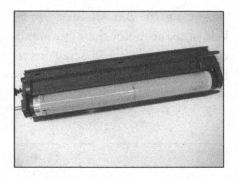

On the opposite side a latch releases the toner cartridge and the drum. Now we're getting seriously messy. Toner is going everywhere.

Running the length of the drum is the corona wire. This fills the drum with negative charges. As the drum rotates, the laser neutralizes the charges at each dot that is to receive ink. After the corona wire sets the negative charges, the laser reverses them so those dots will accumulate the negatively charged toner. The drum rolls around and picks up the toner at those spots where the laser has hit the drum. Paper then rolls against the drum, and the toner is pressed into the paper, in some models assisted by a positively charged roller on the other side of the paper. The paper then rolls to the fuser, which is in the unit labeled "Caution—Heated Section." The heater can reach 400° F!

On the side opposite the laser is a second light path through the black plastic body. This one has a mirror and lens and a photo diode mounted on a circuit board. This is the beam detector that is used to synchronize the laser by sensing it at the end of each pass across the drum.

lens　circuit board

Holding the toner assembly over a trash can, I opened it. Out spilled gobs of toner—but it was interesting. Inside is a paddle wheel that runs the length of the assembly. It pushes toner up, onto a charged roller that holds it against the electrostatic drum. It is turned by a DC motor mounted at one end. A brass gear on the motor shaft drives a plastic gear that turns a worm gear that rotates the paddle wheel.

Along the left side is the large (24 DC volt) motor that drives the rollers pulling paper through. It drives a series of gears, each of which is held onto its axle by a retaining clip. By inserting the tip of a small flathead screwdriver into one of the openings of the clip, you can leverage the clips off (they will go flying) to free the gears.

Two solenoids engage clutches. The solenoids are large coils of wire that magnetically attract a lever arm to disengage the clutch. Each has a small spring to pull the lever arm up to engage the clutch. Although they move only a fraction of an inch, they could lead new lives latching or unlatching a door or box lid.

Back to the middle of the printer, a plastic cover protects a second corona wire. This is a single wire strung across the width of the paper. The wire gives the paper an electrostatic charge so the toner that is held to the drum is pulled off the drum and onto the paper. After the paper passes over this wire and around the drum, it goes to the fuser.

corona wire

The fuser is powered on the right side. An end cap unscrews, and the quartz heating element pulls out. Be careful handling the element as it could break easily into many sharp pieces. Beneath the fuser tube is a roller that is positioned against the tube with springs. In front of the tube is a plastic strip that contains some electronics and a thermostat.

By now, toner is everywhere. Lots more work remains to finish the disassembly, but these are the highlights.

heating element

DPI

Older laser printers delivered 300 dots per inch (dpi). On one of those printers the machine would have to make 90,000 dots to print a one-inch black square. New printers deliver up to 1,200 dpi. These would have to make 1.44 million dots for a one-inch square.

What Now?

This laser printer has a richness of parts that is unmatched in the desktop technology kingdom. The rollers on shafts could make good third wheels on small robots, as long as they are traveling over smooth, flat floors and don't need high ground clearance. The motor is powerful enough to tackle most projects you can come up with, but will require a hefty power supply. The solenoids and microswitches could lock and unlock cabinets or toolboxes. The six-sided mirror could generate some interesting lighting effects. The light-sensitive electrostatic drum begs for experimentation. Can you shine light patterns on it to have it attract toner so you can transfer the patterns to paper? Or, can you create patterns by touching spots with a charged nail?

METAL DETECTOR

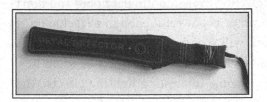

Treasure Cache

   Components on circuit board

   Switches

   Wire

Tools Required

   Flathead screwdriver

   Phillips screwdriver

Yeou can take this to the bank! Or you can take to the bank all the coins you find at the beach. Metal detectors, security wands, and traffic detectors all work the same way. They send out a magnetic pulse and look for a disturbance in the return magnetic field, which indicates the presence of metal.

Lefty Loosens

Three screws on the outside held the two halves of the body together. One more on the inside secured the circuit board to one of the halves. I used a flathead screwdriver to help pry the two body halves apart.

What's inside? Not much! A circuit board with three chips and some other components. One of the chips is a 555 timer chip that can be used in many applications to generate a timed signal, such as turning motors or lights on and off. A piezoelectric speaker also sits on the board. This is the sounding device that alerts you to

the treasure immediately below the detector. When you find the mother lode, the circuit board beeps that piezo speaker.

The on/off switch, electric power leads, and the leads to the coil all plug into the circuit board. I peeled back the yellow tape covering the coil to see the wire windings. The coil is actually two coils in one, hence the four connections it has with the circuit board. Two of the connectors send electric pulses to one of the two overlapping coils to generate a magnetic signal. The other two connectors take the collected magnetic pulse back to the circuit board.

What Now?

It would be fun to mess around with a working metal detector to see if you could make it more sensitive by adding more wire to the coils or changing the size of the coils.

I also wonder if there isn't something useful to be made that could help people locate lost items. Some people always seem to lose car keys or the TV remote. Could you make something to go on a key fob that would be detectable by the metal detector from far away? Could you transform a metal detector into a precious object locator?

One more idea for bike riders: Traffic lights at intersections are often triggered by magnetic detectors in the pavement. They're designed to sense the presence of cars, but a bike may not have enough magnetic material in the frame to trigger the signals. Could the metal detector be transformed to produce a signal large enough to trigger a traffic light yet small enough to be mounted on a bike?

PENCIL SHARPENER (ELECTRIC)

Treasure Cache

AC motor

Detector switch

Gears

Tools Required

¼-inch wrench

Flathead screwdriver

Phillips screwdriver

Wire cutters

Hand-turned sharpeners have two cylindrical cutters that rotate like planets around the pencil. Electric sharpeners have one cylindrical cutter that rotates around the pencil and is powered by a motor. Some are battery powered, but this unit is powered by alternating current.

Lefty Loosens

The question in my mind was *What triggers the motor to start?* It doesn't run until you insert a pencil, but how does it detect that the pencil is there?

I removed the screws from the bottom that held the top and bottom together. Pulling the top off, I could see how the sharpener detected a pencil. A small white plastic slide protrudes into the plastic sleeve where pencils are inserted. As the pencil goes in, it pushes the slide.

The sleeve and the cutter are inside a plastic housing that is held to the motor by two screws. Removing the screws allows it to fall off, exposing the cylindrical cutter. The pencil sits inside the plastic sleeve, and the sleeve and cutting edge revolve around it. At the end of the cylindrical cutter is a small gear that meshes with a plastic ring inside the end of the housing. As the cutter rotates around the pencil, the plastic ring forces it to turn in the opposite rotation to chew away the wood of the pencil. The sleeve and cutter pull off the motor shaft.

cutter

The pencil detector switch is housed in a rectangular plastic body. The top of the body can be pried off with a flathead screwdriver to reveal how the switch works. The plastic slide pushes one strip of copper conductor against the other to complete the circuit and power the motor.

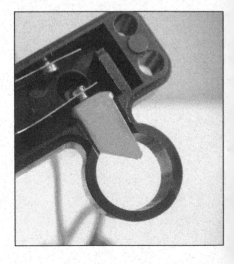

The motor is on the other side of the plastic frame. It is held together by two large bolts, which are more easily removed using a small wrench than a pair of pliers. A plastic bearing holds the rotor in place, and with it removed, the rotor slides out. It has a small plastic helical gear on its shaft. This meshes with a large plastic gear that turns the cutting assembly on the other side of the frame. The small gear has 11 teeth, and the large one has 77 for a 7:1 gear ratio. For every seven rotations of the motor, the sharpener will rotate once.

The motor is called a shaded-pole induction motor. Alternating current moving through the coil of wire that is wrapped around the square metal plates induces a magnetic field. The square plates are called the shading coils. These provide the changing magnetic field to drive the cylindrical rotor.

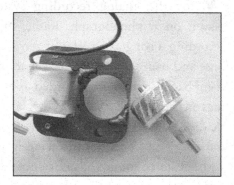

What Now?

If you keep the motor, be careful when using it. The wires are exposed, posing a shock hazard, and the motor has enough torque (turning power) to be dangerous.

The switch is interesting. If you want to use it, I suggest not opening the cover to the switch, as the contacts come out easily and are bothersome to put back in.

PHONOGRAPH

Vinyl records are the storage medium for the analog recording system played on a phonograph. Rather than recording a song as a series of zeros and ones and using digital processing to play them back, sounds are recorded as etchings on the sides of a V-track cut in a plastic disk. The etchings capture the sound intensity and frequency in their wiggles, with left stereo sounds on one side of the V and right sounds on the other. The V-cut follows a spiral track from the outer edge of the plastic record to the inner, just the opposite of CDs and DVDs. A stylus or needle rides in the V-grove to capture the wiggles. Its motion moves a permanent magnet wrapped inside a coil of wire that generates faint electric signals for amplification. You can think of all manner of materials to test under the sensitive needle before taking this apart.

Treasure Cache
  DC motor with pulley
  Drive belt
  Light metal rods
  Microswitch
  RCA plugs
  Rubber feet
  Spindle
  Switch
  Tone arm

Tools Required
  Flathead screwdriver
  Phillips screwdriver
  Wire cutters
  Wrench or pliers

Lefty Loosens

It would have been helpful to first operate the phonograph before dismantling it. But this one was missing its power supply. (It has a 9-volt DC motor but is supplied by household current.)

I removed a clip that held the turntable on the spindle. (The turntable's extra-wide outer edge gives it weight and, with that, more angular momentum to keep the record spinning smoothly.) As I lifted the turntable, the drive belt fell off, and the motor pulley was exposed.

Three screws held the motor on rubber mounts so it could move a bit and when someone hit the spinning turntable the belt wouldn't fly out. I didn't remove the motor right away, as I wanted to look at the underside with the motor in place. Turning the phonograph upside down I removed two of the four rubber feet. Then I removed a few screws that held the bottom cover on.

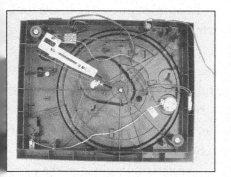

tone arm lift

Beneath the large plastic arm in the upper left corner of the previous photo is the tone arm lift. A lever on the top side of the phonograph allows users to lift the arm so they can place it on the record wherever they want. If you try to lift the tone arm by pulling it up directly, you are likely to drag it across the tracks, scratching the record.

Mounted on the arm is a limit switch to turn the motor on or off, I presume, when the tone arm reaches the end of the record. The spindle is bolted from the underside.

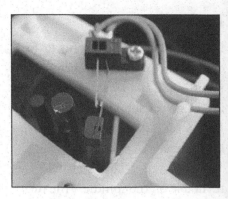

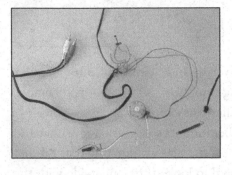

A small circuit board was freed by removing one screw. I expected to see an amplifier on the board, but there was none. The board merely provides connections among the various components. Sounds, as electric impulses, pass through onto the wires leading to the phono or RCA plugs. This phonograph has to plug into an amplifier before going to speakers.

I was most interested in the tone arm. It's held to the base by two screws. Cut the wires and remove the two screws holding the cartridge assembly at the far end. The wires pull out of the hollow arm. (In this phonograph, the stylus or needle is missing.)

The cartridge, glued to the plastic case, requires prying to remove it. Two coils of wire reside beneath a clear plastic cover: one coil for the left sound channel and one for the right.

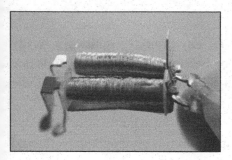

I cut open the plastic cover to get a better look at the coils inside. The wire is very fine.

The Invention of the Phonograph

Thomas Edison's favorite invention was the phonograph (U.S. patent number 227,679). He demonstrated it in 1877 by reciting the poem "Mary Had a Little Lamb" and playing the recording back on his cylindrical record. He created the first talking doll by placing a tiny phonograph inside. The flat disk we know as a record was invented 10 years later by Emile Berliner.

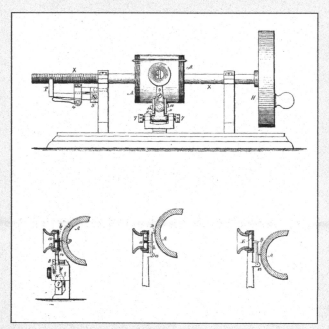

Patent no. 227,679

What Now?

A spinning turntable could be useful for any projects requiring a steady turning motion. Cutting glass bottles by heating them as they turn is one example. Or the turntable itself could become a large wheel. Its spindle could be an axle.

If you are making an electronic instrument—say a guitar—the cartridge could be used as an audio pickup. The motor in this phonograph works and could be powered by a 9-volt battery.

PIANO KEYBOARD

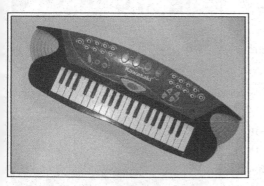

Treasure Cache

    Circuit board

    Plastic keyboards and switches

    Rechargeable batteries

    Recharging jack

    Slide switch and other
       switches

    Speakers

Tools Required

    Flathead screwdriver

    Rotary cutting tool

    Phillips screwdriver

    Wire cutters

These are common finds at thrift stores and garage sales. Inside they are very similar to other electronic musical toys. Note the similarity to the toy guitar (see page 76).

Lefty Loosens

Nine screws held the two halves of
the piano keyboard together. Inside
were two small speakers that lifted
out and several circuit boards.

The keys are interesting; there
are two sets. One has the black keys
made of one piece of plastic, and
the other has white keys made of
another. The black and white keys fit
together and are held in place with
screws. Depressing a key pushes
down on a rubber dome switch
below it. The carbon center inside
the dome touches the circuit board
underneath and connects two sides
of a switch printed on the board.

The piano is a series of switches connected to an integrated circuit that
sends signals to an amplifier and to the speakers. There are more keys
(switches) than wires running from the keyboard to the integrated circuit.
Rather than wire each switch to the integrated circuit, which would require
N+1 wires, the switch contacts are arranged into a keyboard matrix. For
example, if a keyboard had 16 keys, direct wiring would require one output
wire from each plus one input wire (that could connect to each switch) for
a total of 17 wires reaching the integrated circuit. Using a keyboard matrix
and crossbar switch, 4 wires could connect each of 4 keys in 4 rows. Then,
4 more wires could connect 4 keys
in 4 columns. Instead of 16 + 1
wires, only 8 + 1 wires are needed.
The circuitry is able to figure out
which switch in the array is closed
by determining which row and column have current flow.

My piano was powered by rechargeable batteries inside a battery case. They were recharged by inserting an electrical wire into a jack on the piano. The batteries in my device weren't working, but you might find a good set in the piano you take apart.

The Invention of the Electric Piano

The first piano was invented in 1709 in Italy. Synthesizers, first pioneered by telephone co-inventor Elisha Gray in 1876, became popular after Robert Moog created the modern analog synthesizer in the 1960s. Not until 1983 was the first digital keyboard invented that used sampled sounds recorded and stored on a computer chip.

The inspiration for the electronic piano was a bet between the musician Stevie Wonder and prolific inventor Ray Kurzweil. These digital machines provide more accurate representations of a piano but are more expensive. Less expensive versions, like the one we took apart, rely on miniaturized circuits to create the sounds like a synthesizer, rather than playing digital samples.

What Now?

If the piano still works, you can explore the different sounds it's capable of making by bending the circuits. First, make sure that the circuit board is connected to the speakers, and that it's receiving power *only* from the batteries. ***Never bend a circuit that's plugged into alternating current.*** Try connecting different parts of the circuit board with alligator clip leads, other wires, or your fingers, to see if you can generate sounds. With the knowledge that the only thing you might destroy in this process is the remains of an otherwise dead piano, you have the freedom to explore and discover. (See page 79 for more information on circuit bending.)

The piano keys and switch could make a cool-looking switch panel. They could control a bank of LEDs so playing the keys would create visual displays. The speakers could be wired into a radio or other device.

POINTING STICK

Unscrewed Value Index... 4

Cool project rating 4

Treasures to collect 0

Disposal issues 0

Treasure Cache

Empty

Tools Required

Flathead screwdriver

Phillips screwdriver

Scissors

Located in the center of some key-boards, in the middle of letters G, H, and B, the pointing stick allows users to move the computer's cursor on screen merely by pushing. It is an incredibly intuitive device—you use it once and you understand how to operate it.

Lefty Loosens

This pointing stick was riveted to the underside of the keyboard of the laptop taken apart earlier (see page 111). Removing the panel beneath the keyboard and popping the rivets set it free.

The force you apply to the pointing stick, and not its movement, controls the movement of the cursor. This tiny device has two conductors separated by a nonconducting material. As you push the stick in one direction, the force of your finger compresses an electrical strain gauge; the pressure from

your finger pushes the two conductors closer together, making a better connection between them and reducing the resistance of the circuit. An integrated circuit senses the change in resistance and moves the cursor accordingly. Electrical connection is provided by an eight-conductor ribbon cable.

The stick itself is imbedded in the surface of the board. The strain gauges are impossible to see. Even after a layer of circuit board is ground off, all that is visible is a metallic layer. Beneath this, on the underside of the board, is the chip that measures the electric currents coming from the gauges.

What Now?

Take a hike to the trash can and drop in the remains.

The Invention of the Pointing Stick

Edwin Selker came up with the idea for a pointing stick as a way to speed up the operation of a computer. Rather than shifting one hand from the keyboard to a mouse inches away, he wanted to control the cursor with less hand movement. He held that idea for several years until he had the opportunity to develop it. He patented it in 1996 with colleagues at IBM.

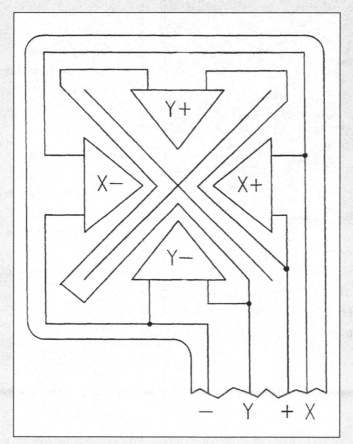

Patent no. 5,489,900

POPCORN POPPER

Treasure Cache
DC motor
Handle
Heating element
Plastic and metal housings
Plastic struts
Switch
Wheels

Tools Required
Alligator clip leads
Flathead screwdriver
9-volt battery
Phillips screwdriver
Pliers
Wire cutters and strippers

It's easy to find popcorn poppers at Goodwill stores. The model I took apart is the most common one. It has some interesting parts and can be rewired into a totally different appliance.

Lefty Loosens

I easily lifted off the plastic top and moved on to the base, where all the interesting stuff is.

The metal cover with the circular opening is held to the plastic frame by four small Phillips screws. Unfortunately, even with these removed and the cover pried up and away from the frame, it doesn't slide off the upward-projecting cylinder. Apparently the cylindrical top was inserted through the opening of the metal cover and glued to the adjoining cylinder. (Finally, to get the metal cover off, I bent the flanged top of the cylinder and pulled it up and off.) The metal cover is a source for strips of light metal.

With the metal cover removed, the working components can come out. They are held in place by two screws in the base. (Initially I missed seeing them, as the screw holes were about the same size as vent holes in the bottom.)

The motor, impeller, and heating element come out together. Fan blades are molded into a plastic disk that rides on the motor shaft. The disk is held tightly on the motor shaft and requires prying with a screwdriver before it slides off the shaft.

Once the wires have been cut, the motor, which is mounted on a small circuit board along with the heating elements and power light, can be pulled out. The circuit board holds four diodes and a capacitor that rectify the incoming alternating current. A diode allows electric current to pass in only one direction. The four diodes constitute a rectifying bridge circuit that converts alternating current into direct current (DC).

The motor is a DC motor, which spins nicely when wired to a 9-volt battery with a pair of alligator clip leads. (I used pliers to break the circuit board holding the motor.) This is a nice motor for some future project.

The heating element has a silver covering that can be pulled off. Underneath are two heating coils.

The outer one is made of larger wire, and it alone is protected by both a thermostat and thermal fuse on the underside of the heating element. The fuse will permanently interrupt the circuit if temperatures exceed 227° C (440° F). This is a bit higher than the popping temperature for corn (about 180° C), probably to ensure no unpopped kernels. My guess is that the smaller inner coil keeps the popcorn warm and the outer coil provides the serious heat to pop the corn.

thermostat   fuse

The power switch comes out the front side of the plastic housing. Its wire connections pull off its contacts. The switch is a useful item.

The plastic legs that support the popcorn popper are held on by long screws. Access these from inside the frame. The legs are screwed together. One leg holds the two wheels, and once the two halves of the leg are separated, the wheels fall out. These plastic legs could become supports or struts.

What Now?

Roasting popcorn and roasting coffee are nearly the same job—but check out the cost for a home coffee roaster and compare it to the cost for a popcorn popper.

Making a coffee roaster from the popcorn popper requires some modifications that involve wiring household voltages (110 volts). (If you aren't comfortable doing this, don't attempt the conversion.) The two electrical components inside the popper, the fan motor and the heating elements, must be separated. Making the fan motor voltage adjustable allows you to control the roasting process and blow out the beans when they reach the roast—light, French, etc.—you prefer.

To make the speed of the fan motor adjustable, purchase a light-dimmer switch. This is the type of switch you often see on the wall in a dining room; you adjust the light level by turning a knob or sliding a lever. Adding the dimmer to the motor circuit allows you to control the roast. Keep the fan motor at low speed until the beans are roasted just right, and then increase the speed by turning the knob to blow the beans up and out of the heater. Since a dimmer switch can be purchased for under $5, you could save $200 by making your own roaster.

RADIO-CONTROLLED CAR

Treasure Cache

Capacitors

Circuit board

DC motors

Gears

Spring

Switch

Wheels, tires, and axles

Tools Required

Flathead screwdriver

Phillips screwdriver

Wire cutters

Radio control was first invented by that techno-genius Nikola Tesla in 1898. But it wasn't until solid-state electronics were invented (the first was the transistor in 1947) that radio control caught on. Today you can purchase radio-control toys for just a few dollars, or buy them by the pound at thrift store outlets.

Lefty Loosens

Three Phillips screws held the plastic body on the frame of this model. As the body lifted off, the wire antenna slid through a hole in the body. There was nothing interesting in the body, so I looked at the frame.

The front of the car is dominated by a circuit board. The antenna and wires to both motors and to the power switch connect to the board. A single screw holds the board in place. Beneath it is the steering mechanism.

Before cutting any of the wires to the circuit board, apply a piece of masking tape to each wire and write on the tape what the wire connects to. This key information will help later in using the circuit board.

Steering is accomplished by a DC motor. You can see the bottom of this motor next to the circuit board. The motor is held in place by a plastic clip. Cut the wires to the motor so you can remove it later, but leave as much wire attached to the motor terminals as possible. This will give you more wire to work with if you use the motor in the future.

On the underside of the car a plastic plate covers the steering mechanism. Removing a few screws allows the plate to fall away; then you can see the pinion gear that sits on the motor shaft on the right side. It drives a large plastic gear, reducing the speed of rotation. This large gear is attached to a smaller gear on top, which drives a gear of much larger diameter. As this last gear turns from side to side, it pushes the steering rod that connects to the two front wheels. Beneath this gear is a small spring that returns the gear to center when the motor is not running. So, the driver merely releases the steering switch, and the front wheels steer straight ahead. Most radio-controlled cars have a small lever you can slide clockwise or counterclockwise to adjust this center position of the steering when the motor isn't running.

electromagnet

The gears slide off metal pins that hold them in place. Turning the car back over, you can push aside the plastic clip holding the steering motor while pushing upward on the pinion gear to remove the motor. Some radio-controlled cars don't use an electric motor to steer them. Some use an electromagnet or a linear actuator. These pull or push the steering rod to one side or the other depending on the direction of current through the circuit and windings.

In the back of the car are the drive motor and gearing, hidden by a plastic cover. To remove it, find four plastic clips that hold a cover over the motor. Many model radio-controlled cars use screws to hold the cover in place.

Rotating the drive motor up allows you to view it along with the gears. On the motor shaft is a small pinion gear that drives a larger plastic gear. This gear has a smaller companion gear that turns another larger gear sitting on the rear axle. These gears reduce the speed of the motor's rotation and increase its torque. The inexpensive motors used in low-end radio-controlled cars run at speeds too fast to operate the cars, so gears are inserted to reduce the speed.

The gears pull out as does the rear axle. The wheels are held on the axle only by friction so tugging and twisting will remove them.

The power or on/off switch is held in place by two small Phillips screws. You can use the switch again, so take a second to remove it.

Radio-controlled cars have many different designs. One has two drive motors and no steering motor. These cars are steered by differential steering: applying more power to the left motor turns the car to the right.

In this model, notice that both motors have a small capacitor connecting the two terminals. These capacitors reduce the electromagnetic radiation noise that the motor generates. Such noise can interfere with the car's reception of control signals from the hand unit. By putting a capacitor across the terminals, high-frequency noise is short-circuited while allowing direct current to flow to the motor.

This model didn't include its handheld control unit. There isn't much of interest inside the hand units except for the switches. If you find a model without the hand unit and you want to operate the car or its circuit board, look on the bottom of the car for the frequency of the control signal—it will be either 27 or 49 megahertz—and then look for a controller with the same frequency. You might have to try several different units (manufacturers use different codes to control their cars) to find one that works with your car, but with some luck you'll locate one.

What Now?

The motors, gears, and wheels are worth more than I paid for this model at a thrift store. The switch is useful in any low-voltage electrical project.

The circuit board may still work, too. If it does and you have the corresponding hand-control unit, you can make a new radio-controlled toy. Connect the circuit board to a battery pack that supplies the same voltage (use the same number of batteries) as the model did. Connect the wires that ran to the original motors to motors or lights or buzzers in your new device.

REMOTE CONTROL (TOY)

Most remote-controlled toys use radio control and are not interesting to take apart. This one uses infrared signals, which is more interesting.

Lefty Loosens

This handheld controller also recharges the toy it controls. It is designed to operate with a small helicopter that stores electric energy in an on-board capacitor. The helicopter has an electric motor but no battery. The capacitor provides power and is lighter than a battery and makes it possible to produce an inexpensive toy. To charge the capacitor, the controller has a bank of six AA batteries. Most radio controllers have a single 9-volt battery.

With the batteries removed, I took out four small Phillips screws that held the two halves of the controller together. The top half lifted off.

Inside the controller are the two control switches, large variable resistors. These control the rotor speed and direction of travel. They are mounted on small circuit boards that are screwed into the lower half of the case. A single screw holds each toggle onto its circuit board.

If you might want to use these as variable resistors, don't take them apart. If you won't use them, remove the screw. Inside is a spring to return the toggle to either its center or its start position.

The circuit board has two small rubber dome contact switches on the lower right side. The plastic cover identifies these as left- and right-spin control adjustment. These control the tail rotor. The two larger switches in the center of the board are the power switch and a three-

position channel switch. By choosing different channels, you can operate this toy without interfering with others using their own identical toy. There are two tiny LEDs, one above and one below the two slide switches. These

LEDs indicate when the helicopter is charging and when the batteries need to be replaced. To see these LEDs mounted on the circuit board I needed a hand lens.

Signals to the helicopter are sent by infrared LEDs mounted on the top of the controller. These are on the small circuit board glued into a transparent plastic dome. I used a flathead screwdriver to pry out the circuit board.

Surprisingly, a fuse is located in the controller. The fuse is wired to one side of the battery pack and appears to protect the three infrared LEDs. They would break in an instant if subjected to more than a couple of volts.

The integrated circuit that controls the helicopter signals is mounted on a small circuit board that sits perpendicular to the main board. Four transistors, a capacitor, and a large resistor are the only other large components.

What Now?

The battery case built into the lower half of the controller is a good find. It delivers 9 volts but uses AA batteries that have much longer life than a 9-volt transistor battery. The toggle switches could make speed or light controllers, but I wasn't able to measure their resistance. The large two- and three-position slide switches required de-soldering to make them available.

If you find this or other remote-controlled toys, grab them. Even if they don't work, they contain a treasure cache of parts. Included in the cache is the capacitor that stores power for the toy.

SANDER

Treasure Cache
AC motor
Bearings
Gears
Metal pad
Springs
Switch

Tools Required
Phillips screwdriver

This orbital sander showed up at my front door. One of my running buddies was holding a moving sale and had been unable to sell it, so he dropped it off on my doorstep with nary a note or a ring of the doorbell. It's cool when good stuff shows up unexpectedly—it's sort of like being visited by the Tooth Fairy, only she brings power tools. Black & Decker must have sold a lot of these, as you see them everywhere. I have my own almost identical sander that thankfully doesn't get much exercise. With a powerful (⅙-horsepower) electric motor inside, this could yield some good parts.

Lefty Loosens

With the sandpaper removed from the bottom pad, I could see two screws in the pad surface. I didn't bother with those right away; instead, I unscrewed the eight or so Phillips screws that held the two metal halves of the case together.

This particular sander had performed a lot of orbits. One of the screws refused to yield, so I attacked it with a rotary cutting tool. Halfway through the screw (having cut into the case above the screw), I tried the screwdriver again, and out came the reluctant screw. The screws holding the two body halves together each had a locking washer to hold them in place while the sander vibrated several thousand times each minute.

Lifting off one side of the case revealed the entire workings.

Electricity passes through a hefty power switch on top and goes to the brushes and the stationary windings. The brushes are housed in two plastic housings on either side of the commutator. The housings slide out of the lower case. As they come out, the springs inside push out the brushes. In operation the springs keep pushing the brushes toward the commutator to make electrical contact. As the brushes wear down, the springs ensure that they continue to make contact.

The top of the motor shaft is held in place by a bearing. The bearing lifts out of the lower case and with it comes the motor.

bearing   commutator

brushes

The rotating part of the motor slides out of the windings. On the motor shaft are the commutator, windings, fan blade, and gear. Another bearing holds the shaft in place just above the gear.

The gear on the motor shaft meshes with a gear on the shaft of an oscillating weight. A bearing holds this shaft in place.

The weight spins around its shaft, but since it is off center, the resulting motion is a vibration. Think of this device as a giant vibrating motor like the one in your mobile phone. In the sander, however, it spins much faster and throws around a much heavier weight.

There are two other things to examine in the sander. One is the screw in the handle that holds a grounding wire to the metal case. Should one of the wires carrying electric power somehow touch the case, the person operating the sander won't get electrocuted. This grounding wire will shunt the power back to the ground in the electrical outlet.

The metal pad that holds the sandpaper pulls out of the case. It has four rubber legs that can be removed by unscrewing the bolts and nuts that hold them on the pad. As they come out, the paper gripper is released, too.

What Now?

The motor could be useful, but probably only if kept in its housing. Without the housing holding both the motor and its bearings, it would be difficult to use.

With everything removed, the bearings and switches have some value. The sanding pad could become the base or frame for another project. With the grippers in place on the pad, you have a nice hand sander.

SCANNER

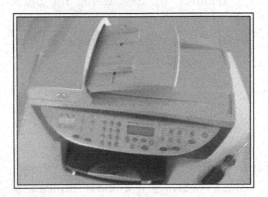

The scanner shown above was a major-league score: a scanner, printer, copier, and fax machine all rolled into one. Even better, it was free, and the owner, Mark, delivered it to me! I think he was anxious to get this out of his house, and I was delighted to have it. This gizmo contains several technologies that collectively have pushed the techno-revolution into hyperdrive. As you hike through this forest of components, you'll see the big timber of a charge-coupled device and a stepper motor amid the ground cover of miscellaneous electronic components and circuit boards.

Treasure Cache
  Charge-coupled device (CCD)
  Cold cathode fluorescent lamp
    and inverter
  DC motors
  Drive belt
  Glass plate
  Lens
  Mirrors
  Stepper motor

Tools Required
  Flathead screwdriver
  Phillips screwdriver
  Wire cutters

LOOK OUT!

Be careful to lift the glass plate out without twisting it. When removing the cold cathode fluorescent lamp, don't bend or twist, as it breaks easily. The mirrors warrant careful handling, too.

Lefty Loosens

To get to the scanning elements I had to remove the paper-feed mechanism that is contained in the hinged door at the top. With that gone, the glass plate lifted out. Beneath all that lay the scan head assembly.

The scan head assembly is where all the scanning action happens. This rides along a stabilizer bar, a bright silver rod that runs the length of the scanner.

The assembly is pulled along the stabilizer bar by a stepper motor. The motor turns a drive belt that clips into the assembly.

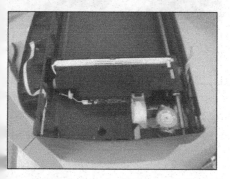

Black covers lift off the scan head assembly to reveal the light source for the scanner: a cold cathode fluorescent lamp. This is the same type of lamp used to illuminate LCD computer screens. Older scanners may have regular fluorescent bulbs, and some have xenon lamps.

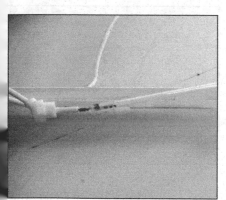

The cold cathode fluorescent is easy to identify, as it is a very thin bulb that reaches across the width of the scanner. And it is easy to break, so be careful. The electrical connection to the bulb is encased in a light, flexible rubber tube.

To get the high alternating-current voltages needed to operate the lamp, an inverter is installed in the assembly. This small circuit board can be identified by the transformer: copper wire windings inside iron plates. The inverter takes the incoming direct-current voltage, converts it into alternating current, and raises the voltage so it's high enough to get the fluorescent bulb to light.

The lamp shines upward to illuminate the page above it. Light that reflects off the page being scanned is bounced from one mirror to another. Depending on the scanner model you tear apart, you may find three or four mirrors. These slide out or are held by metal clips.

The reflected image is collected by a small lens mounted in the center of the scan head. A threaded retaining ring holds the lens in the molded plastic body. Use a flathead screwdriver to unscrew the retaining ring and free the lens.

The lens focuses the image onto a charge-coupled device, or CCD. The CCD used in scanners is a line or one-dimensional CCD—long and thin. The CCDs used in cameras are two-dimensional arrays of sensors and

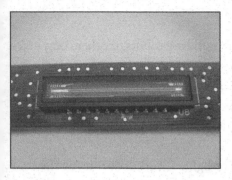

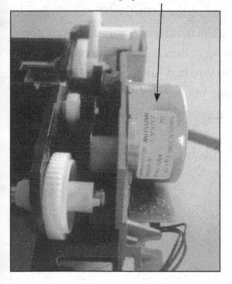

paper feed motor

are square or rectangular (see the digital camera entry, page 14, for a photo of one). The CCD sensor comes off of its companion circuit board, and with some prying the glass cover can be removed. Under 10x magnification, or even 60x, I couldn't see much.

In addition to the stepper motor that pulls the scan assembly along the stabilizer bar, there is a DC motor that drives the paper feed and another motor that positions the printhead along the width of the paper. These are the components of an ink-jet printer (see page 95), so we won't cover them here.

Thanks, Einstein!

Scanners, like digital cameras, take advantage of Einstein's discovery of the photoelectric effect: when light hits some materials, they emit electrons. Light, reflected off your photo or document, hits photosensitive electronic components and generates a flow of electrons that mimics the pattern on the original.

The devices that perform this light-to-electricity sleight of hand are photodiodes. With their associated connections and circuits, they become charged-coupled devices, or CCDs.

What Now?

The stepper motor can be useful if you're building robots or other computer-controlled gadgets. Unlike most direct-current motors, you can't just connect it to a power source to make it go. Look closely at the motor. It doesn't have two electrical connections—it has five or six. A computer has to give commands to move one step at a time, and each step is a fraction of a complete revolution.

Some people have made desk lamps or lighted works of art using the cold cathode lamp and inverter. If this interests you, pay attention to the wiring of the lamp and the electrical connection for the inverter. The lamp should unclip from the inverter. When you have the inverter removed, mount it in a project box and reconnect it to its old power source. You will probably want to insert a power switch in the system so you can turn it on and off. Protect the delicate lamp by inserting it in clear plastic tubing. For more details, search the web for "scanner inverter lamp project." And, be careful of the high voltages!

The lens is fun to fiddle with but is probably too small for most projects. The motors are great. Hold onto them.

SCREWDRIVER (ELECTRIC)

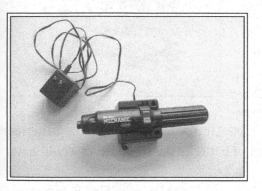

Treasure Cache

Bearing

DC motor

Gears

Plastic body

Power supply

Rocker switch

Tools Required

Allen wrench

Flathead screwdriver

This handy device applies torque to screws without you wearing out your wrist. Its reversible motor lets you screw in and screw out.

Lefty Loosens

Only three Allen screws held the two plastic body halves together. With these removed, the two halves pulled apart at the top, while the bottom remained held together with a plastic clip.

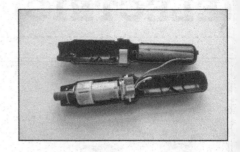

In the top half are two recharge-able batteries. Don't toss these in the trash; they need to be recycled. (In the unit I opened, the batteries were defective, and with a new set of batteries the screwdriver would have worked.)

In the lower half is a hefty motor, electrical switch, locking ring, bearing, and gears. The locking ring is engaged by pressing down on a metal stud that sticks up through the plastic housing. Pushing the stud in engages the locking ring in depressions in the shaft that prevent the shaft from turning.

The electrical switch rocks from side to side to power the screwdriver in clockwise or counterclockwise directions. It also has two terminals that connect to the charging station so that the on-board battery can be recharged. In operation, electric power comes to the switch from the battery. The posi-

tive battery lead connects to one side of the motor when the rocking switch is depressed on the left side. Depressed on the right side, it connects to the other motor terminal. At the same time the switch connects the negative battery lead to the opposite motor terminal.

With the body halves separated, the motor lifts out. It has a plastic mount screwed into the end with the shaft. A small pinion gear on the shaft engages three plastic planetary gears.

A small screwdriver can pry out the three gears. Beneath them is a metal plate that separates these gears from three other gears (made of metal) beneath them. These two sets of three gears spin around inside the metal housing. As the first set spins, driven by the motor, the gears spin a small gear that engages the second, underlying set. These rotate around the metal

housing, and as they do, they drive the output shaft. This arrangement greatly reduces the motor's speed of rotation and increases its torque. The screwdriver's rotation speed is about 100 rpm, determined by counting the number of times it turns. The motor spins many times faster—too fast to count—and there are no markings on the motor to indicate its speed.

To open the recharging base, pry off the bottom with a flathead screwdriver. Inside you can see the metal contacts that connect to the two terminals on the rocker switch inside the screwdriver. That's all there is to the base. The power supply (or wall wart) converts household voltage into 3.5 volts DC.

What Now?

The motors and gears are useable in many projects, as is the switch. The metal housing that holds the gears could be a bearing for an axle or wheel. The plastic housing, with the components removed, could become a handle for something else. The power supply is useful for replacing batteries in other applications.

SHREDDER

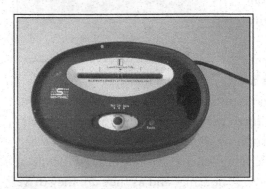

Treasure Cache

    DC motor

    Gears

    Metal rods/axles

    Microswitch

    Shredder rings

Tools Required

    Flathead screwdriver

    Phillips screwdrivers

    Rag

    Wire cutters

P aper shredders have come home—or at least they have shown up in home offices. With a 1984 Supreme Court ruling allowing investigators to search through garbage without a warrant, sales of shredders took off. More recently, identity-theft warnings have pushed sales even higher. Today, old shredders are showing up at thrift stores.

LOOK OUT!

The cutting rings are sharp. (Doesn't that go without saying?) Don't reach in and grab them.

Lefty Loosens

This device had warnings written in several languages instructing users not to open it. One of the holes for the screws that held the two halves of the body together was covered with a sticker warning that the warranty would be voided if the sticker were torn, which would indicate that someone had tried to open the shredder. Shunning the advice, I opened the case anyway.

The only real danger appears to be the possibility of electric shock if the shredder is plugged in while you are working on it. *Make sure you have cut off the power cord, bent the prongs outward, and tossed it safely in the garbage can before you begin disassembling this device.*

One other note on the outside of the shell is a patent number: D 502714. This is a U.S. design patent for the cutting rings that shred the paper. The note also says that patents have been applied for in the United States, Japan, China, and Taiwan.

I removed the five screws that held the two halves of the case together. By also unscrewing the switches from the top, I was able to extract the whole working mechanism, consisting of a forward/reverse switch mounted on a small circuit board, a detector switch to turn the shredder on, a hefty motor, equally hefty gears, and two axles of shredder blades.

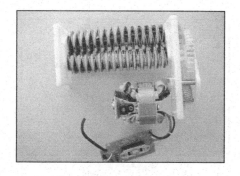

The shredder powers on when you insert a document or credit card into the slot. Some shredders use infrared light detectors to know when to turn on, but this model uses a mechanical microswitch. Pushing your Sears credit card into the slot pushes the arm of the switch, completing the circuit and starting the motor. What's interesting here is the bellows attached to the arm of

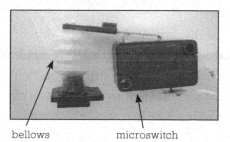

bellows   microswitch

the microswitch. The bellows collapses when the switch is pushed (paper is in the slot). When the paper has cleared the slot, rather than immediately releasing the switch arm, the bellows slowly fills with air, keeping the switch closed for a moment. This allows the cutting rings to continue turning and to spit out any scraps of paper before the shredder motor stops.

Pull the electrical wires out of their connection to the motor. With each removed, a spring underneath pops up, attached to one of the brushes for the motor. Each spring applies pressure to the brushes so they make contact with the commutator on the motor. Interestingly, this is a DC motor but it is running on AC power. The circuit board has diodes that rectify the AC

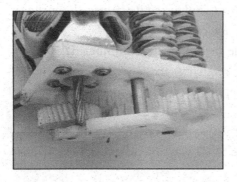

power, but there do not appear to be components to remove the alternating or changing voltage. That is, the voltage that the motor sees will be a positive voltage that varies from zero to 120 volts 120 times each second (for 60-cycle current).

A plastic plate covers the end of the mechanism. Removing it exposes the gear train. (This is greasy work, so grab a rag to clean up.) The motor shaft is a metal helical gear that meshes with a medium-sized gear. The gear train goes from small diameter to large diameter gears to slow the rotation and increase the torque (turning power).

The largest gear drives one of the two shafts holding the shredding blades. It also drives a gear on the second shaft so the two sets of blades rotate in opposite directions.

The two axles holding the blades are screwed into a plastic plate at each end. Removing the plate and a retaining ring allows the blades to come off. The blades are fiercely sharp. They each have small tabs that fit into the axle so they cannot slip.

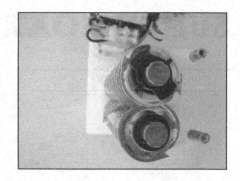

What Now?

The microswitch and bellows grabbed my imagination. Its delay feature for staying on (or off, if you change which of the switch contacts you use) could be useful. For example, if you want a light to stay on for a second after you open or close a door, you could use this switch. I have no idea what the shredder blades could be used for, but the axles and gears hold promise.

SUPER SOAKER

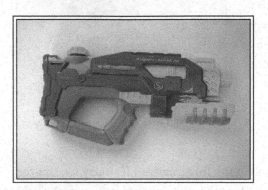

Treasure Cache

    O-rings

    Pump

    Valve with small spring

    Vinyl tubing (short sections)

Tools Required

    Flathead screwdriver

    Phillips screwdriver

    Rotary cutting tool or saw

OK, so this isn't exactly electronic—but it's a cool mechanical take-apart nonetheless. Super Soakers upped the ante in the arms race of squirting water. No more dinky streams of water shot at a range of a few feet. With this pressurized baby you could wallop your neighbor across the street. Independent inventor Lonnie Johnson came up with the new design and later sold his company. Many models show up at garage sales and thrift stores. The one I found has water pistol technology behind the Super Soaker label. The more advanced models have separate chambers for storing water and air pressure.

Lefty Loosens

To open up a Super Soaker, you need to remove the screws and pry open plastic pieces. Removing the screws is easy, although this model had one screw hidden beneath the outermost plastic panel. Even with the screws removed, you need to force the two halves apart. The nozzle cowling also must be removed; this is glued over the two halves of the body. (I cut it off with a rotary cutting tool, but a saw would work well, too.)

With one side of the plastic body removed, the pump is visible. Pulling the pump handle out draws water out of the reservoir. Pushing the handle in forces the water up and out the nozzle. To make this work, there has to be a valve inside the white plastic piping.

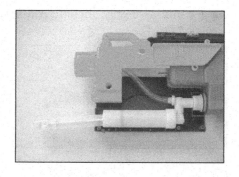

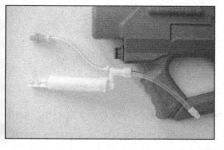

If you want to use the pump, pull it out but don't cut it open. The vinyl tubes are held to the nozzle and water tank by pressure. They pry or pull off. Then the pump assembly comes out.

If your curiosity is greater than your need for a pump, you can dissect it to find the valve. That's what I did.

I used a rotary cutting tool to cut away a section of the plastic tubing. Inside is a spring and rubber stopper. Drawing the handle out pulls the stopper away from the opening, which allows water to flow from the reservoir.

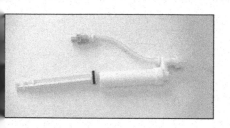

Pushing the handle in forces the stopper up against the plastic housing to prevent water from returning to the reservoir. With nowhere else to go, the pressurized water shoots out the nozzle.

If you score one of the more
expensive Super Soakers with a separate chamber to hold pressurized
air, you will find several valves.

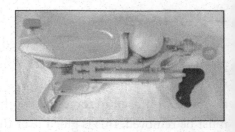

You can guess where they reside
in the several plastic pipes and see
them by cutting the pipe open. Each
valve has a spring and stopper. These models also come with a moveable
trigger and assembly, including a spring and metal pull rod. The air and

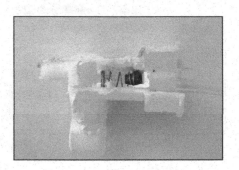

water reservoirs come out easily
once you have the two body halves
apart. But before you extract any of
the components, try to figure out
how the gun works.

What Now?

The components themselves aren't worth much. Two O-rings, a spring, and
some vinyl tubing. The pump probably still works, so figure out a use for it.
Do you need to water hard-to-reach plants?

TIMER

Treasure Cache

AC motor

Tools Required

Flathead screwdriver

Hex or Allen wrench or
    screwdriver

Electric timers—electromechanical timers, that is—show up inside appliances and as timed switches to turn lights on and off while you are out of the house.

Lefty Loosens

This timer has cogs or pins that you set to trip the switch and turn a light on or off. Two sets of cogs are provided for the faceplate shown here: one color turns on the light, and the other turns it off. The lamp plugs into the timer, and the timer plugs into an electrical outlet. The dial turns slowly, one revolution per day, and carries the cogs past the switch. The cogs flip the switch to open or close electrical contacts to control power for the lamp.

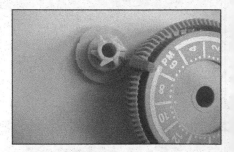

The faceplate dial turns in only one direction, clockwise. You set the time of day by rotating the dial to the correct position.

The faceplate can be pried off to reveal the plastic gear that turns the plate. At the upper left corner is the switch that is turned when a cog pushes against it. The switch also has a rotating knob for the user to control the lamp manually.

Looking closely at the underside of the dial or faceplate shows the ratchet system that allows the dial to turn only clockwise. Turning it clockwise allows the three plastic tabs to click over the ridges on the inside of the dial. Pushing in the opposite direction forces the tabs into the ridges, preventing the dial from rotating.

The timer is held together by two hex head screws. Inside is a gear-motor assembly and switches.

Copper bands carry electric power from the timer's prongs that are inserted into an outlet to the receptacle mounted on the side of the timer. The ends of two bands are pushed together by the switch or allowed to spring apart (circuit open).

The gear motor is an AC synchronous motor, which means its speed of rotation is determined by the frequency of the alternating current that drives it. The power company maintains the frequency at or near 60 cycles per second in the United States, so the motor spins at that rate. Since the dial

turns once per day and the motor spins 60 times a second, gears are needed to slow down the rotation rate. (The metal gearbox defied my meager attempts to open it further.)

What's an Electromechanical Timer?

This device is called an electromechanical timer. Other timers available include mechanical and electronic timers. Mechanical timers operate without electric power; you twist the dial to both set the time and provide the energy to operate the timer. Electronic timers provide a digital read-out on an LCD screen. These are powered by a small battery and use the oscillations of a quartz crystal, like many watches, for keeping time.

What Now?

You could modify a working timer to act as a 24-hour clock or a motor.

TOOTHBRUSH (ELECTRIC)

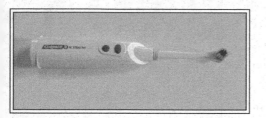

Treasure Cache

Brush

DC motor

Metal shaft

Metal case (motor)

Switch

Water-resistant housing

Tools Required

Alligator clip leads

Batteries

Flathead screwdriver

Rotary cutting tool

D eveloped in the 1950s to help people with limited mobility brush their teeth, electric toothbrushes have expanded their market reach. The one shown below operates on two AA batteries. Other units have recharging stations for NiCad batteries or have inductive charging stands to prevent shock hazards on the bathroom countertop. This model oscillates back and forth, but others vibrate or use ultrasonic sound waves instead of brushes.

Lefty Loosens

Remove the brush by pushing on the tab on the back side of the housing behind the brush. If the brush doesn't fall out, slide the screwdriver underneath the round plastic base of the brush and pry it out. Underneath you can see the crooked end of the motor shaft. A metal spindle beneath the brush keeps it in place and allows it to rotate.

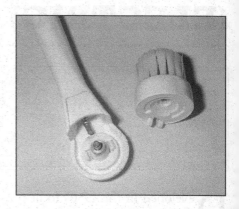

While in operation, the motor spins in one direction, but the brush oscillates back and forth about 50 degrees. The crooked end of the motor shaft acts as a crank. It transforms the circular motion of the motor into back-and-forth motion in the brush.

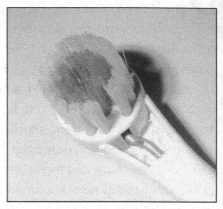

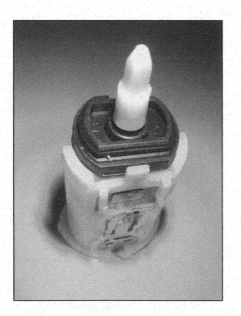

The plastic housing for the motor shaft pulls out of the plastic case that holds the motor and batteries. A small latch holds it in place, but you can pull it out with a yank. With the batteries and brush removed, you can try to pull the motor assembly out of the battery case. The motor goes into the brush this way and latches into position. If the latch is weak, you may be able to yank it out; otherwise you will have to cut open the case.

Use a fine-toothed saw or rotary cutting tool to cut around the perimeter of the plastic case, just below the power buttons. Peel back the case to release the motor.

Check out the switch mounted into the plastic case holding the motor. Pressing down where the "On" button would hit causes the end of the switch to make contact. This also buckles the center band of the switch that holds the contact down. Pressing down on the buckled band releases its hold and allows the switch to open. This switch is a nice find for any applications you might have for a small motor.

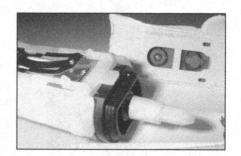

What Now?

The motor encased in the water-resistant plastic case would make a great project motor for any in-water models. A submarine model, for example, would be a great use of the motor, switch, and housing.

Test the motor operation with new batteries. If it works, cut the plastic housing near the brush end to expose the metal shaft. Check the back of the shaft for a metal clip that holds the shaft in place and cut above this. You will probably need to cut off the end of the metal shaft because it is bent to fit into the head of the brush.

If the motor does not respond to new batteries, it still might work. Remove the motor from the plastic case as outlined above and test it with alligator clip leads supplying electricity from two 1.5-volt cells in series or one 9-volt battery.

TOUCHPAD

Treasure Cache

    Grid

Tools Required

    Phillips screwdriver

    Scissors

    Wire cutters

Magically your fingers move the cursor across the screen—remote control with no obvious mechanism. Pretty cool. These are common on laptops and netbooks, taking the place of a mouse, since airport-gate waiting areas offer neither the flat surfaces nor the room to spread out that a mouse requires. And with no moving parts, a touchpad doesn't have much that can break.

Lefty Loosens

The touchpad is like the other components in a laptop or other computer device: it comes out as a completely separate component. It is screwed into the laptop frame and connects with a small wiring harness. A small Phillips screwdriver allows you to remove it.

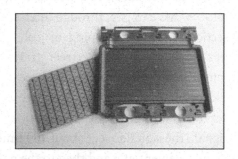

This touchpad has four buttons. Each makes contact with a circuit on the plastic film below it to register your intentions.

Freed of its retaining screws the touchpad assembly comes out easily. The circuit board has a chip that connects to columns along the top and rows along the left side.

Flipping the board over shows the arrangement of connections. Columns are spaced 1 centimeter apart.

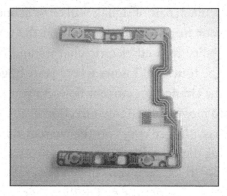

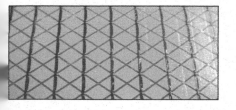

The method by which the touchpad knows where your pinky is pointing is called capacitance sensing. As your finger wanders over the pad, even without touching the pad, the electric charges on your finger change the distribution of charges in the pad, and these changes are sensed by the circuit. The pad has many lanes and avenues crisscrossing, so any given point where the distribution changes can be related to an intersection of streets.

Think of a city with streets inter-secting at right angles. Only in this case, the north-south streets are all raised 10 feet above east-west ave-nues so cars can pass through inter-sections without stopping or col-liding. Now shrink this street grid into a small rectangle about three inches by four inches, and replace cars with a high-frequency electric

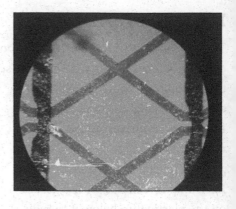

signal. The signal is sent about 100 times each second to one street and each avenue sequentially and then to the next street and each avenue.

Each pair of signals will cross at one unique intersection. When you place your finger at one location on the touchpad, it changes that intersection's ability to hold electric charges and changes the current passing through. The touchpad knows where your finger is at each point in time by sensing the changes in current flow. As you move your finger, it directs the cursor to move in the same direction at the same speed. The center of your finger has a larger effect on the current than the side of your finger; by compar-ing the capacitance at adjacent intersections, the touchpad can be more pre-cise in locating your finger. One manufacturer claims that its model has an expected lifetime of a million taps.

What Now?

The treasure cache is nearly empty on this project. You do get a cool-looking grid, but we haven't figured out what to do with it. Maybe you could stick it on a bulletin board and ask people if they can identify it.

The Invention of the Touchpad

When George Gerpheide switched to an Apple Macintosh, he discovered how difficult it was to constantly move his fingers off the keyboard and then back again to switch between typing and using the mouse. Although he recognized that the graphical user interface of the Mac was the way to go, he found that the back-and-forth hand movement was inefficient. So he started thinking of ways to beat the mouse.

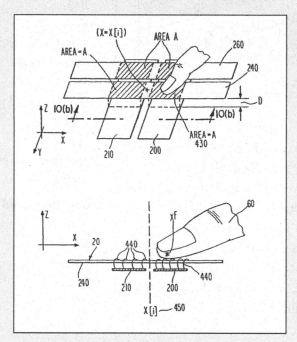

Patent no. 5,305,017

He later told me that he numbered his prototypes Cat #1, Cat #2, and so on, as they were alternatives to mice. By Cat #19, he had discarded his starting technology (using resistors in a membrane) in favor of using capacitors. Cat #20 was a model that worked well enough to sell, so he launched Cirque Corporation, which he later sold. Today he continues to invent in his basement.

TYPEWRITER

I bought this portable model—complete with plastic case—for pennies per pound at the Goodwill Outlet store near me. Any resemblance to a modern keyboard is purely superficial. This baby is heavy, and it has hundreds and hundreds of parts. If you have ever doubted the intelligence of our species, take one of these apart. Somebody had to be pretty clever to design this thing.

Treasure Cache
Bell
Keys
Metal rods, struts, bars
Pulley/wheels
Ribbon reels
Rubber-coated rollers
Screws
Spools
Spring-loaded pulley
Springs

Tools Required
Flathead screwdrivers
Paper clip
Pliers
Wrench

Lefty Loosens

This is not a 10-minute take-apart. There are many dozens of screws awaiting you, some of which present difficult access. There are many dozens of springs, metal linkages, and levers as well.

I started by removing the plastic cover of this manual typewriter. A few flathead screws underneath held it onto a solid metal frame. It takes some twisting and wrestling to get the cover off, but once removed, the inner workings are revealed.

Next you can remove the typing ribbon and the two spools that hold it. The spools could make wheels or pulleys for a lightweight project. The ribbon is surprisingly strong material. (Although pretty messy with ink, this ribbon had dried out, so my fingers weren't turned black.) Cut the ribbon and pull it out. The spools should lift out as well.

I popped off the plastic letters on the keys of my typewriter; they were held on by friction only, so a pair of pliers and a wrist twist got them off.

You might glue a piece of refrigerator magnet (cut to size with scissors) to the back of each letter key so you can leave notes or write (short) poems on your refrigerator door.

Next up is the carriage and platen, the large roller that you turn to feed paper into the typewriter. After removing the plastic covers off each end, search below these to find brackets holding the platen assembly in place. The roller itself could make a great wheel for some device.

A spring-loaded pulley attached to the platen assembly pulls the platen assembly across the page while you type. It unscrews and begs for some creative use. Other pulleys are mounted on the metal housing to guide the cord pulling the platen.

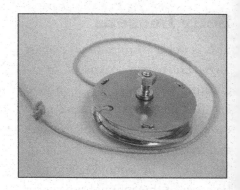

Behind the carriage is the tab bar. It contains many, many metal pegs, one for each letter position on a page. The typist sets the tabs by pushing a peg into position. The individual pegs are held in place in a metal bar.

Removing the top of the bar frees up the pegs, and out they come. These would be great for demonstrating an electromagnet, but they must have other uses too.

Beneath the platen assembly are other roller bars that guide paper into position. These pop out of their brackets.

More challenging is removing the keys that strike the ribbon and their linkages. The keys are contained in what is called a basket, and they pivot on curved, metal rods that have to be extracted from the basket. Play with the

keys to find where they pivot. The curved metal rods are at the pivot points. The ends may protrude from the housing, and you can pull these out with pliers. If the end doesn't stick out, insert a small nail or paper clip to push out the rod, then grab the other end with pliers and pull.

Each letter key has two springs and several linkages. It will take a few minutes to extract them all.

Typewriters have bells to alert the typist that he or she has reached the end of the line and must push the carriage back. As the carriage moves with each typed letter, a tab gets closer to hitting the bell, just in time for you to type a few more letters before hitting the absolute end of the line. (The bell in this model was riveted onto the metal frame. The only way to get it out is to cut the frame away. Other models may make it easier to salvage this ringer.)

If you find an electric typewriter rather than a manual, you will recover an electric motor. Later electrics—the IBM Selectric, for example—used a type wheel or type ball rather than individual keys to strike the ribbon. To spin the type ball and get it into position requires metal ribbons or belts and pulleys driven by a motor. Have fun!

What Now?

The spring-loaded pulley is the most intriguing piece here. It would power a spring-powered model car or some other device. Pull out the end of the string to wind the spring, and then release both the spring and the model and watch it go. If you are work-ing on an electric typewriter and the motor still works, it could have uses as a rotary cutting tool. Keep in mind, however, that anything mounted on the motor shaft has to be *anchored absolutely* to prevent fly-off and injury. Use the keys to spell out messages.

UNINTERRUPTIBLE POWER SUPPLY (UPS)

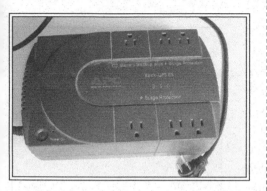

Treasure Cache

   Aluminum heat sinks

   Ferrite chokes

   Piezo speaker

   Wires

Tools Required

   Flathead screwdriver

   Phillips screwdriver (long handle)

   Rotary cutting tool

   Wire cutters

As your safety net for an electrical supply that suffers from fluctuations in voltage, brownouts, or blackouts, a UPS can save you huge expense in lost software and damaged hardware. A UPS keeps going even when the electric power does not. It also isolates computers and sensitive drives from the spikes and valleys that occur in line voltage. When power goes out, the UPS regenerates the 120-volt alternating current that the computer needs by drawing on its internal battery. The battery supplies direct current, so the circuit board includes an inverter, which creates alternating current with a smooth sine-wave form current. This alternating current passes through the transformer to increase its voltage up to 120 volts, and is fed to the plugs mounted on the outside of the UPS case.

Lefty Loosens

This puppy is heavy. Most electronic components and peripherals are light-weights, but not the UPS. The reason for its heaviness is a lead-acid battery and a huge transformer.

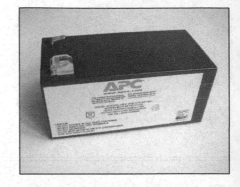

Step one was to remove the battery from the housing. A single Phillips screw held the plastic flap in place. With the screw removed, the flap slid off, and the battery lifted out—a 12-volt battery, like the one in your car, only smaller. The battery in this system is rated to provide power for six hours. After that, you'd be in the dark.

All the other components are in the top half of the housing. It's easier to remove them if you first cut the electrical wires connecting them.

The biggest and heaviest component in this half is a large transformer. This slides out of the plastic case. Cut or pull the wires off their connectors to remove it.

The circuit board has a few interesting components. A piezo speaker sits in the middle of the board. If you want to go to the trouble of de-soldering it, you could retrieve this (probably) working speaker. Also on the board is a smaller transformer.

Adjacent to the transformer on the circuit board are two power transistors. The power transistors are easy to recognize: they have three stout legs and a black body. Power transistors are used to switch moderately high currents. These are screwed to large blocks of aluminum, which are heat sinks that draw heat away from the transistors. Removing the screw lets you pry the heat sinks up and off the board using a flathead screwdriver.

A circuit reset button protrudes out of one side of the plastic case. It is encased in two pieces of black plastic held together by three rivets. (Unable to pry it open, I cut two of the rivets with a rotary cutting tool.) Inside is a piece of metal that holds the button in and keeps the power on. When the metal overheats, it bends upward, allowing a spring to push the button out. The user has to reset it to get the UPS to work again.

Power cords coming into the UPS and leaving it are wrapped around and through two ferrite toroid chokes (see page 70). These metal donuts reduce electromagnetic radiation that may interfere with the electronic devices being powered by the UPS.

The Invention of the UPS

The inverter circuit that lies at the heart of UPS systems was invented in 1973 by Bruce Wilkinson (U.S. patent number 3,769,571). This includes a system to detect power failures and immediately start supplying power from a battery.

What Now?

The aluminum heat sinks could become weights, strong supports, or even bearings.

If the battery holds a charge, it could be useful. You would need a charger that could deliver the voltage the battery requires.

From the circuit board you can de-solder the piezo speaker for use in anything that requires a beep. The piezo can play a range of frequencies, but the sound level will vary greatly depending on the frequency. At mid-ranges, a few hundred to thousand cycles per second, the sound should be strong. Above or below that, the sound level will diminish sharply.

VIDEOCASSETTE CAMERA

T his is a wonderful find. Hours of technology delving and a bucketload of cool parts await you.

Lefty Loosens

This disassembly started with medium Phillips screws and progressively worked down to tiny Phillips screws. A handful of screws held the two halves of the plastic body together. The metal frame inside was screwed to the right half of the plastic body and easily came out. On the left side is a drive belt connecting a motor to the capstans that move the film through the camera. On the right side is the scissors assembly that lifts out to accept a cassette tape and pushes in to insert the tape into the tape drive.

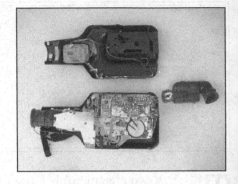

Before digging deeper into the metal frame and its contents, I removed the eyepiece. Everything is held together by Phillips screws, and no prying or forcing is required to separate the parts.

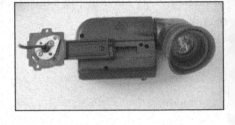

Mounted on the eyepiece is a microphone for recording the sound.

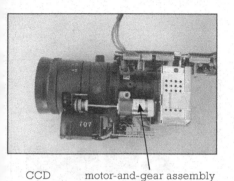

CCD   motor-and-gear assembly

Inside the eyepiece are several lenses and mirrors. The eyepiece shows the operator what the camera sees by projecting a video image on a tiny video viewer. This long, skinny vacuum tube, along with its electronics, is mounted inside the eyepiece. They show the image on a screen near the viewer's eye.

The lens's plastic sides come off easily to reveal a motor-and-gear assembly on each side of the lens. By turning the gears I saw that one motor provides the focus and the other drives the zoom in and out. The latter responds to the "W" (wide) and "T" (telephoto) button switches. Both motors spin when I connect them to a 9-volt battery.

Mounted on a small circuit board at the back end of the lens—which I disassembled by removing a dozen very small screws—is the CCD. This converts the photons of light from sunlight reflecting off Grand Teton into electronic signals for processing by the circuit board above the lens so you can show off your vacation.

Beneath the lens are two identical-looking lenses mounted in square plastic cases. The assembly holding both lenses is held on by screws. Behind the lens is a small circuit board that is attached with screws. (I used a flathead screwdriver to pry the left lens out.) Behind the lens is an electronics sensor, which is a light detector. Behind the other lens is an infrared LED. The camera determines distance to the subject for autofocus by sending out an infrared beam and measuring the time required for it to reflect off the subject and

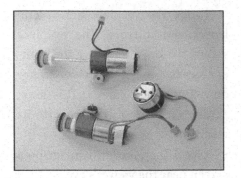

return to the sensor. If the subject is far away, the signal never returns and the camera focuses on infinity. Both lenses magnify two to three times.

The lenses that focus light into the camera are inside the lens. Two lenses are held in a cylinder with slots cut into its side. The focus motor rotates the cylinder and that brings these two lenses closer or moves them farther away.

Also inside the lens is one more motor. This one operates the lens opening or aperture. The motor rotates a tiny set of arms that pull two pieces of black paper together or apart to control the amount of light that enters.

I haven't gotten to the tape moving and recording components of

the camera yet, but since much of the rest of the camera closely mimics the videocassette recorder (see page 207), I will cover only the major features.

Like the VHS player, the camera uses two sides of the metal frame to move and process the tape. Tapes go into the scissors jack on one side; this side has two capstans that fit into the cassette to spin the feed and take up reels. One of the capstans has a brake to prevent the film from going too fast and getting tangled. The rollers that move the tape onto the spinning head are on this side, as is the spinning head, the large silvery cylinder set at an odd angle. The tray that holds the tape moves out to accept the tape and in to set it in place so the rollers can pull it out of the cassette and onto the head. Remove the screws on the top of the spinning head to lift the top off and see inside.

Also like the videocassette recorder, the camera has two additional fixed heads. One, advertised on the outside of the camera as the "flying erase head," is for on-the-fly erasing. The other is the audio head that reads and records audio signals. These signals are recorded on a different part of the tape than is the video signal and don't require being read by a fast-spinning head.

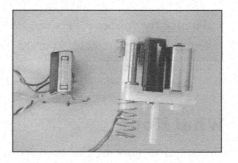

The motor for moving the tape itself is mounted on the other side of the frame. This motor, a DC brushless motor, has a pulley mounted on top. Freed from beneath, the motor pulls apart. Inside are the coils of wire (poles on the stator) that create the magnetic fields that drive the motor. Inside the top is a circular magnet (the rotor) that the induced magnetic field pushes against. The pulley on top drives a belt that connects to another pulley centered above the cassette tape.

pulley   magnet

coils

Also on this side are the gears that move the tape in toward the spinning head and the mechanism. To move the gears, there is a small DC motor with belt drive to a gear train. The gears turn a worm gear that turns the large assembly of gears.

Finally, the rubber rollers can be lifted or pried off their axles. The gear assembly will pry apart to separate the gears. Springs can be removed. This is a treasure trove of parts.

What Now?

The small motors in the lens are not the inexpensive, fast-turning DC motors. They move at a sedate speed and with little torque, making them uniquely useful. The lenses are fun to experiment with. The scissors assembly that lifts a cassette into and out of the camera could become a storage device to hold small objects, maybe a radio, and keep it out of the way until you need to access it. The eyepiece slides along a track, and this could provide a base for moving. The trick in each case would be to remove the mechanism from its frame without damaging it.

VIDEOCASSETTE RECORDER (VCR)

This machine caught on very quickly in the 1970s and is disappearing almost as quickly now that it has been replaced by the DVD player. Old VCRs show up at thrift stores quite often. Taking a VCR apart offers active entertainment for about as long as the average movie.

Unscrewed Value Index..9+

Cool project rating........5+

Treasures to collect.........4

Disposal issues 0

Treasure Cache

DC motors

Drive belts and pulleys

Fuse

Gears

LCD

LED

Magnets

Metal housing

Rollers

Screws

Springs

Switches

Tools Required

Allen wrench (minor use)

Flathead screwdriver

Phillips screwdriver

Wire cutters

Lefty Loosens

A few Phillips screws held the bottom panel on this machine. The protective metal housing was held on only by a few clips. Depressing them with a flathead screwdriver released it. I held onto the housing; it could be useful if cut into strips or other shapes.

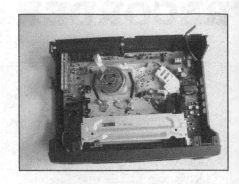

The front cover also came off by pushing down a few plastic tabs, revealing the user control switches, the LCD, an LED, and an infrared receiver (to the right of the LCD) that receives the signals from a handheld remote.

The metal assembly that moves the tape is held by a few Phillips screws, and with them removed, the assembly lifts out.

The front part of this assembly holds the VCR cassette and moves the tape down onto two posts that pull the tape out and hold it against the spinning head. These posts slide through slots in the metal frame. The spinning head, the silver-colored cylinder, sits behind the metal frame that holds the cassette.

Removing the screws that hold the upper metal part of the assembly will allow you to see the rest better. It is impressive to think of how many individual parts there are in a VCR, and that each one required design, manufacture, and installation.

In addition to the spinning head that reads the video signal, there are two stationary heads on opposite sides of the tape-handling assembly. Cutting one open reveals a coil of wire inside. Since this wire extends over the

full width of a tape, I assume this is a quick-pass erase head. The other stationary head reads the audio signal that is magnetized at the top of the tape.

The mechanism that moves the cassette and pulls the tape in and out of it is a complex system of plastic gears and arms. It is powered by a small DC motor. The motor is in a

plastic case that is held in place by a single screw. The case itself unclips to release the motor. A flathead screwdriver can be used to pry off the circuit board that provides power to the motor. The motor has a worm gear on its shaft. I touched the motor terminals to a 9-volt battery and was pleased to see it spin.

Two black capstans fit into the bottom of a cassette and rotate the tape and take up reels. One of them has a brake pad that can be engaged to keep it from spinning too fast.

Next up is the spinning head held to the metal assembly by three screws underneath. The brushless motor that drives the spinning head sits on top of the head. The top of the motor is a circuit board that has two screws that hold it onto a metal disk below. The underside has the poles of the motor, the wrappings of copper wire. These spin around inside

capstan

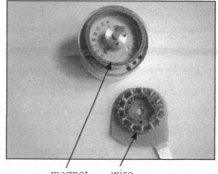

magnet   wire

the circular magnet that remains on the shaft.

Removing this disk requires a tiny Allen wrench. Once the Allen screw is loosened, the top of the spinning head lifts off to reveal circular windings of wire inside.

A corresponding set of circular windings is in the lower, spinning half. Separating it from the circular magnet of the motor required taking out two more screws. The spinning head is held stationary at the bottom, and the top doesn't move either. Only the middle spins.

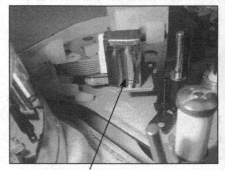

stationary head

One more motor awaits. This is another brushless motor, and its job is to pull the tape along. It is located underneath the metal assembly. On top of the motor is a pulley with a belt that drives a capstan. The motor unscrews from the top of the assembly. The stator poles are screwed to a bearing that allows the motor shaft to spin while the stator remains stationary.

Underneath the tape-handling assembly is a large circuit board that fills the base of the VCR. The circuit board is held in place by several plastic tabs, which can be pried out of the way with a flathead screwdriver to lift it out.

At the back of the circuit board is a bright metal cage that is held together by clips. Prying it open reveals a circuit board with induction coils on one

side and other components on the
other. The cage protects this circuitry that brings in a video/audio
signal from an outside source (an
antenna or cable connection) and
sends out the video/audio signal to
a television.

Most electrical appliances don't have their own fuses, but the VCR does.
The fuse is located just where the electrical power cord joins the circuit
board.

Whew! That's a lot of components, a lot of screws, and a lot of work. Now
you have a good pile of parts.

What About the Videotape?

The tape inside a videocassette is 800 feet long, and it moves
through a VCR at a speed of about 1.8 inches per second. If the
tape moved past a stationary read head at this speed, it would
take 330 cassettes to hold one movie. Since a single cassette
holds a movie, we can be sure there is some creative engineering going on inside.

The revolutionary idea that made VCRs work was to not use a
stationary head, but to spin the head past the tape. Although the
tape moves slowly, the head spins past the tape at 25 miles per
hour.

What Now?

The motor and gears can power a model car or boat. The rollers can become
wheels. The metal case provides material for a chassis. The brushless motor
stators would make a nice decoration. What else can you think of?

VIDEO GAME
LIGHT GUN

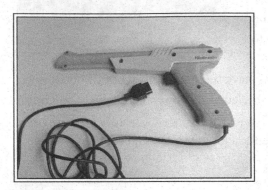

These show up at garage sales, thrift stores, and online auctions. This is an ingenious device for getting information (specifically, where you are aiming the gun) into the game system to see if you are accurately aiming at the target. Creating this magic requires an assortment of components that you will find useful for other projects.

Treasure Cache
- Lens
- Metal weights
- Microswitch
- Photo detector
- Springs

Tools Required
- Phillips screwdriver
- Scissors or wire cutters

Lefty Loosens

A half dozen or so Phillips screws held the two halves of the gun together. I removed them and lifted off the left half of the plastic shell.

lens   weight

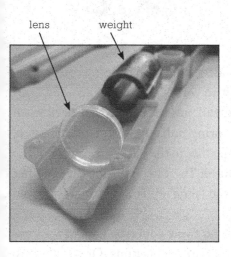

photo detector   shield

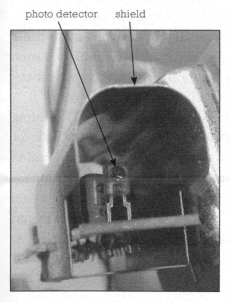

Near the end of the barrel is a small, clear plastic lens that keeps dust out of the gun and focuses light on the photo detector (a photodiode) behind it. The lens probably won't start any fires even in the brightest sun, but it does magnify images better than the magnifying lens on a Swiss Army knife.

The next object in the barrel is a hollow weight. It's hollow so that light can pass through it. This weight, and the shiny weight in the handle, give the gun the right feel. Remove them and hold the gun to feel how different—i.e., wimpy—it is.

Behind the hollow weight is a shield that covers the photo detector and a small circuit board. This is wired to the trigger assembly, which is below and behind it. The photo detector and shield pull out; the two halves of the gun are all that hold them in place.

Three screws hold the trigger assembly in place. Several smaller screws hold the cover on the assembly. Beneath the cover are three springs, the trigger, and a microswitch. As you pull the trigger, you compress the larger spring at the bottom. Releasing your trigger grip allows this spring to push the trigger back.

microswitch

Also, as you pull the trigger, the plastic slide along the top pushes the microswitch arm; this closes the circuit so your shot can be recorded. This action also elongates the top spring that will pull the plastic slide back when the plastic catch releases it. The third spring pulls up on the plastic catch, keeping it in place. As the trigger slides farther to the back, the catch will slip off, allowing the top slide to move forward. The catch will slip off just after the microswitch has closed. You hear the "ping" of the switch closing and opening as you pull the trigger.

The microswitch has the lever arm and three contacts. One of the contacts is common and is used in either of two applications. One of the two remaining switches is normally open (NO), so using this contact and the common allows the switch to close the circuit. If the other contact, normally closed (NC), is used with the common, the circuit is closed until the switch lever is depressed. We use these as touch sensors for robots. They have many other uses, so keep this one and keep the springs.

What you won't find inside this gun is a light source or light-emitting diode. In other words, this gun doesn't shoot. Instead, the photo detector detects a blast of white light on the video screen that surrounds the position of the target you are aiming for. Pulling the trigger and closing the microswitch causes the screen to go blank for one frame, $\frac{1}{30}$ of a second, and then to highlight the target for one frame. If the optical sensor inside the gun "sees" the blast of light during the second frame, the computer scores it as a hit.

What Now?

The microswitch can be used in two different modes. In the video game light gun, it completes a circuit when the trigger is depressed, so you can use it to turn on a motor or light. But you can also use it to turn *off* a motor or light. There are three contacts on the switch: "NC" stamped on the side stands for normally closed. "NO" is for normally open. The third contact is common. Using the "NC" and common contacts, you can use the switch to make a model turn off its motor when it runs into a wall.

The photo detector warrants some investigation. I suggest removing the metal shield so you can get to the contacts. Clip onto them and, using a voltmeter, see if a bright light allows current to flow.

WEB CAM

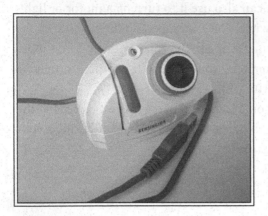

Treasure Cache

Active pixel sensor

Electrical cable

Lens

Tools Required

Phillips screwdriver

Wire cutters

T hese tiny video cameras are inexpensive yet work wonderfully well. Plug one into your computer, and you're ready to start shooting.

Lefty Loosens

A few Phillips screws held the back of this web camera to the front. With the back removed, the front half lifted off the cylindrical post that was part of the base. The front half has a circuit board that is held in by two screws.

Taking out the circuit board allows the lens to come out of the plastic shell. The lens focuses light onto the large chip with the glass covering.

Underneath the cover is an active pixel sensor. Like a charge-coupled device (CCD), an active pixel sensor (APS) converts captured light at each pixel into electric charge that can be measured to give a digital value. An APS has an amplifier integrated into the circuit for each photodiode or light detector. Prior to the development of active pixel sensors, CCDs were the chip of choice for converting optical images into electronic signals. CCDs are still widely used in video and still cameras. Because active pixel sensors have built-in amplifiers, they can

acquire images faster and with less power than CCDs can. Their lower cost favors them in many applications, especially low-end ones.

Eric Fossum, Sunetra Mendis, and Sabrina Kemeny invented the CMOS active pixel sensor (patent number 5,471,515) in 1995.

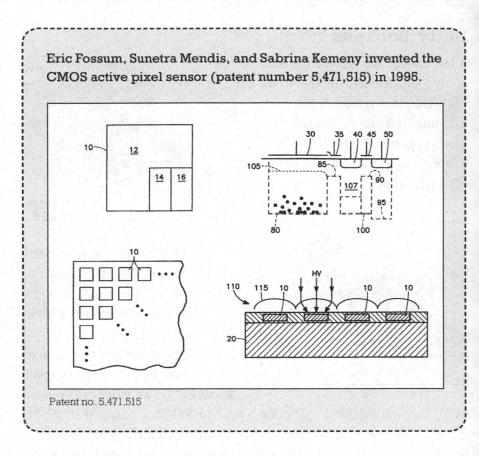

Patent no. 5,471,515

What Now?

The lens could have a variety of uses, from starting a fire when you run out of matches (with its small lens, you'd better be at low latitudes on a clear day) to helping you read the menu when you forget your glasses. The APS is cool to look at—techno eye candy—so could be of use in some decoration.

WIRELESS ROUTER

Treasure Cache
   Antenna
   Light guides
   Plastic case

Tools Required
   Phillips screwdriver
   Scissors

One day last summer I was sitting in my tent at a campground in the middle of Montana, checking my e-mail with my laptop. Some may be horrified by the idea that technology has crept into the great outdoors, but I was delighted that I could sit outside on a perfect summer day and, unfettered by wires, converse with the world.

Lefty Loosens

First check out the back of the router. You'll see the connector for the power cable, a reset button, and the jacks for the incoming Internet signal and outgoing connections to computers—should you not want to use the wireless function. The power switch is located here, as are two antennas.

Inside are clear plastic light guides that direct the light from tiny LEDs up to the top of the plastic case so that you can be mesmerized by the blinking lights.

light guides

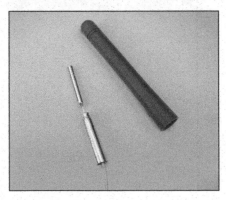

The chips on the circuit board perform several functions. Several store data or directions. Others generate the radio signals that are broadcast. A metal cover shields some of the circuitry from radio waves. Leads for the two antenna stretch across the circuit board.

Inside the plastic antenna shell is a metal tube attached to the antenna wire. That's it.

So how does a wireless router work? An onboard radio generates electric signals that travel to the antenna. The antenna converts these signals into electromagnetic waves that radiate through space to the wireless card in your computer.

The radio signals are transmitted either at 2.4 gigahertz or 5.0 gigahertz, where gigahertz means a billion cycles per second. For comparison, your favorite radio talk show personality shouts at you at 1610 kilohertz, a carrier wave that oscillates one millionth as fast. The high frequency for wireless systems allows much higher rates of data transfer but limits the distance the signals can be transmitted.

The Invention of the Wireless Router

Norman Abramson at the University of Hawaii built the first wireless computer network that allowed computers on several islands to communicate with each other in 1970 (U.S. patent number 6,108,314).

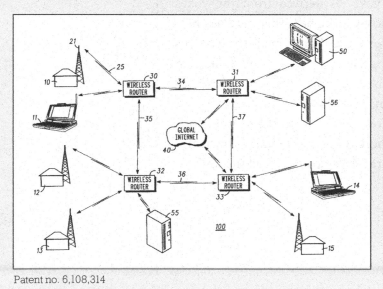

Patent no. 6,108,314

What Now?

I suggest attaching the two antennae to your favorite baseball cap to stir up some conversation.

YOKE

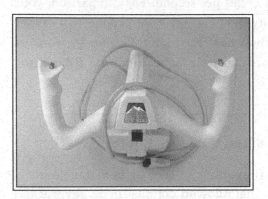

Treasure Cache

Potentiometers

Spring

Switches

Tools Required

Flathead screwdriver

Phillips screwdriver

Used for computer games, the term "yoke" is derived from the yokes used to fly airplanes. This yoke has buttons for each thumb, plus the obvious steering action, left and right. Like an airplane yoke, this one moves in and out to command pitch movements (nose up and nose down) in airplane simulators. One more control is the throttle, which is a separate lever to one side of the steering wheel.

Lefty Loosens

Ideal for reverse engineering, this yoke came apart with the removal of a few screws. The two thumb switches are momentary contact switches: they pass electric current only as long as they are depressed. The three other controls are connected to potentiometers or variable resistors.

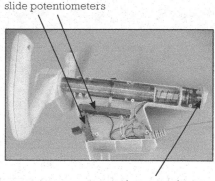

slide potentiometers

rotating potentiometer

The steering motion effected by turning the wheel twists a potentiometer at the distant end of the steering column. Turning the wheel one direction increases the resistance in a circuit, and turning the other way decreases it. The resistance of the potentiometer ranges from zero to about 90,000 ohms—a value I got from my multimeter.

rotating potentiometer

rotating potentiometer

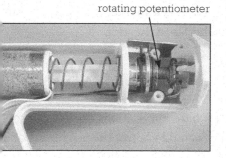

The other two controls move slide potentiometers. Rather than rotate to change the value of a variable resistor, these slide back and forth. The throttle control is attached to a resistor mounted on the inside of the plastic control panel. The pitch control potentiometer rides in a slot cut around the metal steering column so the operator can control the pitch regardless of what angle the steering wheel is turned. As the operator pushes the yoke in and out, the pitch control potentiometer changes its resistance.

slide potentiometer

slide potentiometer

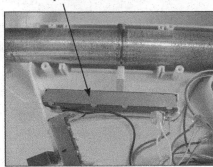
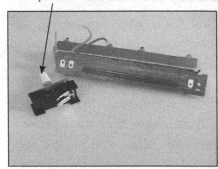

The fact that there are no circuit boards inside the yoke suggests that the game software interprets the current flowing through the five different circuits to know what the operator is doing. The computer supplies +5 volts to each circuit, and the potentiometers in each circuit (except for the two momentary contact switches) control the current passing through that circuit. High resistance gives lower current flow.

Presumably, the steering column could have been made of plastic to save some money. I assume its metal construction is a marketing touch, not an engineering one: a heavier yoke may feel better to the user.

What Now?

The potentiometers, spring, and switches have applications for lots of projects. For low-voltage applications, the pots could control the brightness of a light or the speed of a motor.

The Invention of the Yoke

The first yoke intended for use with microcomputers was invented by Robert J. Kuster and Jay B. Ross, who assigned rights to New Flite, Inc. Their invention (patent number 4,659,313, issued in 1987) does not look or work like the more modern one I described.

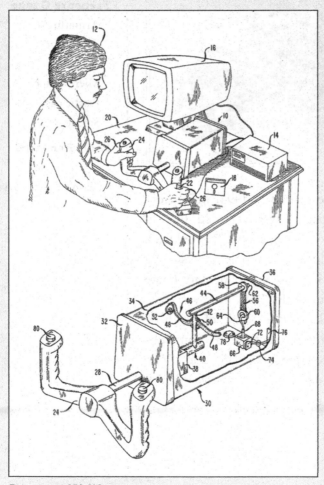

Patent no. 4,659,313

ZIP DRIVE

Treasure Cache

   Armature

   DC motor

   Magnets

   Microswitch

   Plastic case

   Solenoid

Tools Required

   Allen wrenches

   Flathead screwdriver

O n the market in 1994, well before the patents were awarded, the Zip drive quickly became a hit. It provided a great solution to the problems of storing and retrieving large sets of data for home and office users. By 1998, however, technical problems—the "click of death"—destroyed the product's reputation. Subsequent developments of CDs and DVDs for storing data assumed the role that Zip drives had briefly filled.

Lefty Loosens

Zip drives work like floppy drives, but they store data more densely and can retrieve it more quickly. Zip drives use a higher-quality coating on the disks that allows data to be stored more densely. Also, the read/write head is smaller, which allows more tracks to be written on a disk. Unlike a floppy drive, the armature that moves the heads in and out works like a hard drive. Floppy drives heads are moved by a small electric motor, while Zip drives and hard drives use voice coil solenoids. These move in and out rapidly as electric currents induce magnetic fields in the coils surrounding the armature. The electromagnetic fields push and pull against permanent magnets in the drive to move the arm.

This take-apart wasn't as easy as it could be. The two halves of the body were held together by tabs instead of screws. I forced a flathead screwdriver into each of the slots along the midline and twisted the driver. Opening two on one side allowed me to lift the top piece away from the bottom.

That was only half of the battle. With the plastic case removed, there was still a metal case surrounding the workings. With a screwdriver I leveraged this open. Inside this model the screws were all small with five-sided heads, or were Allen screws.

The motor assembly sits in front. On top of it is a yellow plastic arm that moves to the rear when a disk is inserted. This activates a switch so the drive knows that it's time to go to work. The motor assembly is held loosely in place by two springs. These allow it to slide along two side rails. It lifts out, and beneath it is the circuit board with memory and processing chips. The on/off switch that projects out of the front plate of the drive pushes on a tiny microswitch on the circuit board.

The arm that moves the heads in and out is located under a piece of clear plastic. This can be lifted off once the Allen screws are removed. Attached to the arm is a rectangular coil of wires surrounding bar magnets. The position of the arm is controlled by changing the magnetic field in the coils. The other end of the arm has a pair of tiny heads that magnetically read the contents of the disk.

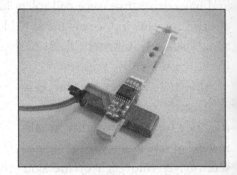

Off to one side is a solenoid. This solenoid is a linear actuator; passing an electric current through it either pushes the plunger out or pulls it in. Office and apartment doors with electric locks use similar solenoids or actuators.

solenoid

What Now?

The solenoid is pretty cool. Make an electronic pinball game using the solenoid to launch tiny steel balls (bearings). Or use it to whack on a bell when you want attention.

The Invention of the Zip Drive

Zip drives are protected by some twelve patents. The first was granted in 1996 to John Briggs and David Jones for the linear actuator that positions the read/write heads on the disk (patent number 5,508,864).

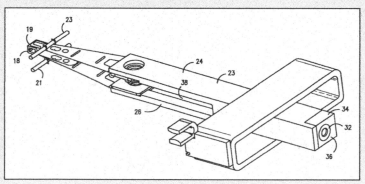

Patent no. 5,508,864

GLOSSARY AND INDEX OF COMPONENTS

AC motor. AC, or alternating current, is the electric energy you get by plugging something into a wall outlet. (Batteries, on the other hand, produce DC, or direct current.) With AC power, the electrical voltage oscillates 60 times each second. In the United States it varies from −120 volts to +120 volts. A motor powered by alternating current has a wide variety of uses, but only if you are able to *safely* connect it to a 120-volt outlet. Slow-rotating motors can operate pulley systems, signs, kinetic sculptures, etc. Fast-spinning motors can drill or chill (as in a fan). **135, 162, 181**

Active pixel sensor. A device for detecting and recording photo images. It consists of an array of light sensors, each of which has an adjacent amplifier that makes it "active." Active pixel sensors are used in cell phones and web cams. They also look cool, and can be turned into high-tech decorations. **216**

Aluminum heat sinks. Heat sinks draw heat away from sensitive components, and as air circulates through and around them, they dissipate the heat. Aluminum sinks can be reshaped or cut or used whole as weights, supports, or bearings. **197**

Antenna. Metal rods that capture electromagnetic waves, such as radio waves, and convert them into electric signals. They also work in reverse, taking electric signals and converting them into electromagnetic waves. The telescoping variety can become pointers and extendable prods for reaching that item in the back of the cabinet. **56, 219**

Armature. A support that holds other components. (The term is also used to describe the rotating part of an electric motor.) **86, 226**

Ball (mouse). 32

Battery case (AA). The compartment that holds a device's batteries. It's quite useful for projects requiring battery power—it makes it much easier to connect power to the project. **76, 158**

Battery terminal (9-volt). The connector that snaps onto a 9-volt battery to provide power to a device. It can be reused to connect power to homemade devices as well. **29**

Bearing. A device that allows an axle or other component to move, but only in a particular direction. **162, 171**

Bell. 192

The Way Toys Work
The Science Behind the Magic 8 Ball, Etch A Sketch, Boomerang, and More

Ed Sobey and Woody Sobey

A Selection of the Scientific American Book Club

Profiling 50 of the world's most popular play-things—including their history, trivia, and the technology involved—this guide uncovers the hidden science of toys. Discover how an Etch A Sketch writes on its gray screen, why a boomerang returns after it is thrown, and how an RC car responds to a remote control. This entertaining and informative reference also features do-it-yourself experiments and tips on reverse engineering old toys to observe their interior mechanics, and even provides pointers on how to build your own toys using only recycled materials and a little ingenuity.

978-1-55652-745-6
$14.95 (CAN $16.95)

The Way Kitchens Work
The Science Behind the Microwave, Teflon Pan, Garbage Disposal, and More

Ed Sobey

If you've ever wondered how a microwave heats food, why aluminum foil is shiny on one side and dull on the other, or whether it is better to use cold or hot water in a garbage disposal, now you'll have your answers. *The Way Kitchens Work* explains the technology, history, and trivia behind 55 common appliances and utensils, with patent blueprints and photos of the "guts" of each device. You'll also learn interesting side stories, such as how the waffle iron played a role in the success of Nike, and why socialite Josephine Cochran *really* invented the dishwasher in 1885.

978-1-56976-281-3
$14.95 (CAN $16.95)

A Field Guide to Household Technology

Ed Sobey

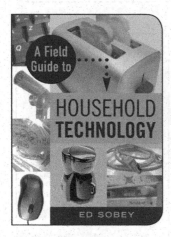

978-1-55652-670-1

$14.95 (CAN $18.95)

Illustrating how a bathroom scale measures body weight and what the "plasma" is in a plasma-screen television, this fascinating handbook explains how everyday household devices operate. More than 180 different devices are covered, including gadgets unique to apartment buildings and houseboats. Devices are grouped according to their "habitats"—living room, family room, den, bedroom, kitchen, bathroom, and basement—with detailed descriptions of what they do and how they work, and photographs for easy identification. You'll never look at that pile of remote controls on your coffee table the same way again.

A Field Guide to Office Technology

Ed Sobey

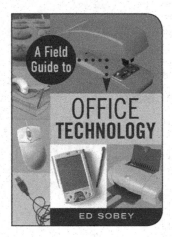

978-1-55652-696-1

$14.95 (CAN $18.95)

The modern office can be a confusing technological landscape. How does a motion detector spot intruders? What is a voltage surge, and why does a computer need to be protected from it? And why do telephone keypads have their 1s in their *upper* left corners, while calculator keypads have their 1s in their *lower* left corners? Entries for more than 160 devices tell you how they work, who invented them, and how their designs have changed over the years. No longer will you need the IT staff to explain that mysterious blinking box in the coat closet.

A Field Guide to Roadside Technology

Ed Sobey

"Fun, informative, and easy to use."
—*School Library Journal*

If you've surveyed the modern landscape, you've no doubt wondered what all those towers, utility poles, antennas, and other strange, unnatural devices actually do. In *A Field Guide to Roadside Technology*, more than 150 devices are grouped according to their "habitats"—along highways and roads, near airports, on utility towers, and more—and each includes a clear photo to make recognition easy. Once the "species" is identified, the entry will tell you its "behavior"—what it does—and how it works, in detail.

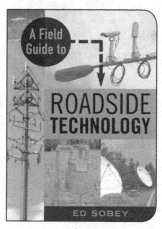

978-1-55652-609-1
$14.95 (CAN $20.95)

A Field Guide to Automotive Technology

Ed Sobey

If you don't know your catalytic converter from your universal joint, *A Field Guide to Automotive Technology* is for you. How does an airbag know when to deploy? What is rack and pinion steering? And where exactly does a dipstick dip? More than 120 mechanical devices are explored in detail, including their invention, function, and technical peculiarities. You'll also find information about components found on buses, motorcycles, bicycles, and more. Even seasoned gearheads will learn from this guide as it traces the history and development of mechanisms they may take for granted.

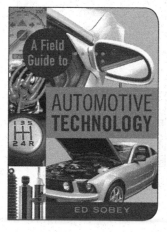

978-1-55652-812-5
$14.95 (CAN $16.95)

978-1-55652-779-1
$16.95 (CAN $18.95)

Haywired
Pointless (Yet Awesome) Projects for the Electronically Inclined

Mike Rigsby

Written for budding electronics hobbyists, *Haywired* proves that science can inspire odd contraptions. Create a Mona Lisa that smiles even wider when you approach it. Learn how to build and record a talking alarm or craft your own talking greeting card. Construct a no-battery electric car toy that uses a super capacitor, or a flashlight that can be charged in minutes then shine for 24 hours. Each project is described in step-by-step detail with photographs and circuit diagrams, and helpful hints are provided on soldering, wire wrapping, and multimeter use.

978-1-55652-520-9
$16.95 (CAN $18.95)

Gonzo Gizmos
Projects & Devices to Channel Your Inner Geek

Simon Field

This book for workbench warriors and grown-up geeks features step-by-step instructions for building more than 30 fascinating devices. Detailed illustrations and diagrams explain how to construct a simple radio with a soldering iron, a few basic circuits and three shiny pennies; how to create a rotary steam engine in just 15 minutes with a candle, a soda can, and a length of copper tubing; and how to use optics to roast a hot dog, using just a flexible plastic mirror, a wooden box, a little algebra, and a sunny day. Also included are experiments most science teachers probably never demonstrated, such a magnets that levitate in midair, metals that melt in hot water, and lasers that transmit radio signals.

Available at your favorite bookstore, by calling (800) 888-4741, or at www.chicagoreviewpress.com

CHICAGO
REVIEW
PRESS